The Technology of Discovery

The Technology of Discovery

Radioisotope Thermoelectric Generators and
Thermoelectric Technologies for Space Exploration

Edited by

David Friedrich Woerner
Jet Propulsion Laboratory
California Institute of Technology, USA

Registered Office
John Wiley & Sons, Inc., 111 River Street, Hoboken, NJ 07030, USA

Editorial Office
9600 Garsington Road, Oxford, OX4 2DQ, UK

For details of our global editorial offices, customer services, and more information about Wiley products visit us at www.wiley.com.

Wiley also publishes its books in a variety of electronic formats and by print-on-demand. Some content that appears in standard print versions of this book may not be available in other formats.

Library of Congress Cataloging-in-Publication Data
Names: Woerner, David Friedrich, editor.
Title: The technology of discovery : radioisotope thermoelectric generators and thermoelectric technologies for space exploration / David Frederich Woerner.
Description: Hoboken, NJ : Wiley, 2023. | Includes bibliographical references and index.
Identifiers: LCCN 2022051673 (print) | LCCN 2022051674 (ebook) | ISBN 9781119811367 (hardback) | ISBN 9781119811374 (adobe pdf) | ISBN 9781119811381 (epub)
Subjects: LCSH: Thermoelectric generators. | Radioisotopes in astronautics. | Thermoelectric apparatus and appliances. | Outer space–Exploration.
Classification: LCC TK2950 .W647 2023 (print) | LCC TK2950 (ebook) | DDC 621.31/243–dc23/eng/20221108
LC record available at https://lccn.loc.gov/2022051673
LC ebook record available at https://lccn.loc.gov/2022051674

Cover Design: Wiley
Cover Images: © NASA/JPL-Caltech/Malin Space Science Systems

Set in 9.5/12.5pt STIXTwoText by Straive, Pondicherry, India

Contents

Foreward

I am pleased to commend the Jet Propulsion Laboratory (JPL) Space Science and Technology Series, and to congratulate and thank the authors for contributing their time to these publications. It is not easy for busy scientists and engineers who face constant launch date and deadline pressures to find the time to tell others clearly and in detail how they solve important and difficult problems. So, I applaud the authors of this series for the time and care they devoted to documenting their contributions to the adventure of space exploration. In writing these books, these authors are truly living up to JPL's core value of openness.

JPL has been NASA's primary center for robotic planetary and deep-space exploration since the Laboratory launched the nation's first satellite, Explorer 1, in 1958. In the years since this first success, JPL has sent spacecraft to each of the planets, studied our own planet in wavelengths from radar to visible, and observed the universe from radio to cosmic ray frequencies. Even more exciting missions are planned for the next decades in all these planetary and astronomical studies, and these future missions must be enabled by advanced technology that will be reported in this series. The JPL Deep Space Communications and Navigation book series captures the fundamentals and accomplishments of those two related disciplines. This companion Science and Technology Series expands the scope of those earlier publications to include other space science, engineering, and technology fields in which JPL has made important contributions.

I look forward to seeing many important achievements captured in these books.

Laurie Leshin, Director
Jet Propulsion Laboratory
California Institute of Technology

Note From the Series Editor

This title is the latest contribution to the Jet Propulsion Laboratory (JPL) Space Science and Technology Series. This series is a companion series of the ongoing Deep Space Communications and Navigation Systems (DESCANSO) Series and includes disciplines beyond communications and navigation. DESCANSO is a Center of Excellence formed in 1998 by the National Aeronautics and Space Administration (NASA) at JPL, which is managed under contract by the California Institute of Technology.

The JPL Space Science and Technology series, authored by scientists and engineers with many years of experience in their fields, lays a foundation for innovation by sharing state-of-the-art knowledge, fundamental principles and practices, and lessons learned in key technologies and science disciplines. We would like to thank the support of the Interplanetary Network Directorate at JPL for their encouragement and support of this series.

Jon Hamkins, Editor-in-Chief
JPL Space Science and and Technology Series
Jet Propulsion Laboratory
California Institute of Technology

Preface

"Far better is it to dare mighty things, to win glorious triumphs, even though checkered by failure . . . than to rank with those poor spirits who neither enjoy nor suffer much, because they live in a gray twilight that knows not victory nor defeat."

— *Theodore Roosevelt*

Radioisotope Thermoelectric Generators (RTGs) have proven to be one of the most creative solutions to the challenge of providing a reliable source of electrical power for deep space missions. RTGs have powered spacecraft that have discovered Ocean Worlds in our solar system, evidence that Mars harbored persistent liquid water in its ancient past, the largest known glacier in the solar system on Pluto, that Enceladus—a moon of Saturn—spouts icy plumes fed from a subterranean ocean, and given us our first taste of interstellar space.

This book is a history, primer, reference book, and partial record of what has become one of the most successful technologies ever fielded for the sake of space science, and an educated look at potential futures for RTGs. This title binds all of this together with an explanation of the basic physics of how RTGs are able to produce power from heat, along with discussions of specific programs, missions, failures, and more.

The history of RTG production has been episodic. The first RTG was invented during the Eisenhower administration in the 1950s and gave rise to frequent RTG-powered space missions in the 1960s and 1970s. However, in subsequent decades, the pace of missions that required this enabling technology waned to approximately the one to two launches per decade we see today. Production of various types of RTGs was stopped and restarted in concert with demand. This led to a dwindling of the industrial base and expertise required to produce these unique generators, and imprinted an ebb and flow or episodic pattern onto the production rate of RTGs over time.

Today, the US Department of Energy and NASA have formed a strong team using a "constant rate" approach to serve the demand and needs of a power-hungry community of space explorers. RTGs provide power where the sun does not shine regularly or strongly enough, and, we believe, RTGs will continue to be a vital power and energy source for complex deep space missions for generations to come. They are solid-state, quiet, extremely reliable, and long-lived.

The history of RTGs is glorious. That may sound like hyperbole, it is not. Some RTGs have been powering spacecraft in flight for over 45 years. No RTG has ever caused a mission or spacecraft failure. Their performance and lifetimes are unmatched by any other power source for spaceflight that humankind has mastered. They require no periodic maintenance in flight. RTGs are electrical generators that enable humankind to explore our darkest, dustiest, coldest regions of the solar system, and the vastness of the space outside of our sun's minuscule heliosphere.

RTGs use materials that exhibit the Seebeck effect to convert heat into Direct Current (DC) electrical power. The Seebeck effect was first reported in 1823 by the Baltic German physicist Thomas Johann Seebeck. His discovery is the foundation for the technologies employed in RTGs for space missions using thermoelectric couples.

RTGs convert heat to power by directing heat flow through thermoelectric couples. Heated thermoelectric couples produce a voltage and when a load is applied, current will flow from each couple. Voila, DC electricity. Yet, RTGs are products of chemistry, materials science, nuclear physics, and decades of sophisticated engineering. They are solid-state, have no moving parts, and each is unitary, needing no ancillary equipment to produce power. Since their creation, extensive and well-funded research, engineering, and flight experience forged modern RTG designs for space exploration.

RTGs are not simple. They are annular in design and employ rings of thermoelectric couples arrayed around a stack of heat sources to convert heat to electricity. Drawings of them tend to unintentionally mask some of the internal components, some of which can reach temperatures over 1,000 degrees Centigrade or 1,832 degrees Fahrenheit, and are maintained at those temperatures for years. It is difficult to comprehend what happens to materials at those temperatures. Chemical reactions are vastly accelerated over our day-to-day experiences. Materials that are solid at room temperature may sublimate when heated to such temperatures. In addition, materials can diffuse into one another.

A schematic of a thermoelectric couple suggests simplicity and masks several phenomena, such as the dramatic temperature reduction between its hot- and cold-side components. For example, Figure 0.1 shows a silicon-germanium thermoelectric couple (aka unicouple), which comprises five subassemblies that

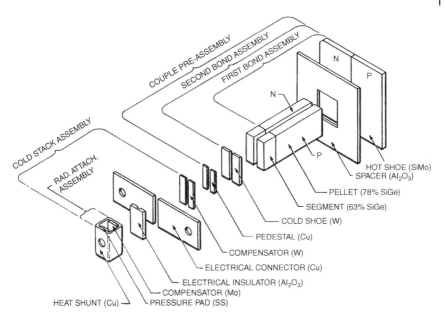

Figure 0.1 A schematic of a silicon-germanium thermoelectric unicouple used in the GPHS-RTGs flown on the Cassini mission. Parenthetical letters refer to chemical elements or compounds. Credit: Lockheed Martin.

include eighteen separate components. During steady-state operation in today's RTGs, the temperature along the length of this type of unicouple drops from 1,035 °C (1,895 °F) at the hot shoe to 290 °C (554 °F) at the heat shunt. That gradient stretches just over one inch and yields a temperature change of ~750 °C (~1,382 °F). This dramatic temperature gradient suggests deep, unobvious complexities tied to thermoelectric technologies and RTGs.

Insight into the unobvious is what you will discover in these pages and should provide the reader with enough RTG history to make them conversant, enough flight history to incorporate this knowledge into deep space mission designs, some background to inform policy and funding decisions, and comprehensive insights intended to help sidestep many of the failures of the past. We employed real-world examples throughout and applied theory to practical situations. We intend this book to bridge the gap between university studies and professional work. Therefore, we provide extensive lists of references to point readers to fundamental and referential sources.

The material herein complements and expands the public history of RTGs, explores the current state-of-the-art, and attempts to peer forward in time. We intend to provide more historical detail of both successes and failures, discuss advances and setbacks of novel thermoelectric materials and technologies that scientists have investigated and abandoned, and chart possible courses to future RTGs and their technologies.

The reinvigorated search for life beyond Earth and NASA's drive to install a permanent human presence on the Moon have breathed new life into space exploration, and each thrust will require technologies that operate effectively in cold and dark spaces. RTGs provide hope for these immediate goals and will enable many more important discoveries. It is a marvelous time to be exploring space!

David Friedrich Woerner,
Jet Propulsion Laboratory/California Institute of Technology,
Pasadena, California, USA

Authors

Brian K. Bairstow
Jet Propulsion Laboratory/California
Institute of Technology
Pasadena California

Chadwick D. Barklay
University of Dayton Research
Institute, Dayton, Ohio

Russell Bennett
Teledyne Energy Systems, Inc.
Hunt Valley, Maryland

Thierry Caillat
Jet Propulsion Laboratory/California
Institute of Technology
Pasadena, California

Ike C. Chi
Jet Propulsion Laboratory/California
Institute of Technology
Pasadena, California

Emory D. Collins
Oak Ridge National Laboratory
Oak Ridge, Tennessee

Shad E. Davis
Idaho National Laboratory
Idaho Falls, Idaho

David W. DePaoli
Oak Ridge National Laboratory
Oak Ridge, Tennessee

Patrick E. Frye
Aerojet Rocketdyne
Canoga Park, California

Nidia C. Gallego
Oak Ridge National Laboratory
Oak Ridge, Tennessee

Lawrence H. Heilbronn
University of Tennessee
Knoxville, Tennessee

Tim Holgate
John Hopkins University Applied
Physics Laboratory, Laurel, Maryland

Chris L. Jensen
Oak Ridge National Laboratory
Oak Ridge, Tennessee

Steve Keyser
Teledyne Energy Systems, Inc.
Hunt Valley, Maryland

Andrew M. Lane
Aerojet Rocketdyne
Canoga Park, California

Young H. Lee
Jet Propulsion Laboratory/California
Institute of Technology
Pasadena, California

Jong-Ah Paik
Jet Propulsion Laboratory/California
Institute of Technology
Pasadena, California

Kaara K. Patton
Oak Ridge National Laboratory
Oak Ridge, Tennessee

Brian Phan
Jet Propulsion Laboratory/California
Institute of Technology
Pasadena, California

Stan Pinkowski
Jet Propulsion Laboratory/California
Institute of Technology
Pasadena, California

Glenn R. Romanoski
Oak Ridge National Laboratory
Oak Ridge, Tennessee

Kevin L. Smith
Jet Propulsion Laboratory/California
Institute of Technology
Pasadena, California

Michael B.R. Smith
Oak Ridge National Laboratory
Oak Ridge, Tennessee

Ying Song
Teledyne Energy Systems, Inc.
Hunt Valley, Maryland

George B. Ulrich
Oak Ridge National Laboratory
Oak Ridge, Tennessee

Joe VanderVeer
Teledyne Energy Systems, Inc.
Hunt Valley, Maryland

Hsin Wang
Oak Ridge National Laboratory,
Oak Ridge, Tennessee

Karl A. Wefers
Aerojet Rocketdyne
Canoga Park, California

Robert M. Wham
Oak Ridge National Laboratory
Oak Ridge, Tennessee

Christofer E. Whiting
University of Dayton Research Institute
Dayton, Ohio

David Friedrich Woerner
Jet Propulsion Laboratory/California
Institute of Technology
Pasadena, California

Andrew J. Zillmer
Idaho National Laboratory
Idaho Falls, Idaho

Reviewers

Chadwick D. Barklay
University of Dayton Research
Institute, Dayton, Ohio

Charles E. Benson
Jet Propulsion Laboratory/California
Institute of Technology
Pasadena, California

Thierry Caillat
Jet Propulsion Laboratory/California
Institute of Technology
Pasadena, California

Eric S. Clarke
Idaho National Laboratory
Idaho Falls, Idaho

Joe C. Giglio
Idaho National Laboratory
Idaho Falls, Idaho

Terry J. Hendricks
Retired—Jet Propulsion Laboratory/
California Institute of Technology
Pasadena, California

Douglas M. Isbell
Jet Propulsion Laboratory/California
Institute of Technology
Pasadena, California

Stephen G. Johnson
Idaho National Laboratory
Idaho Falls, Idaho

Vladimir Jovovic
Jet Propulsion Laboratory/California
Institute of Technology
Pasadena, California

Emily F. Klonicki
Jet Propulsion Laboratory/California
Institute of Technology
Pasadena, California

Christopher S. Matthes
Jet Propulsion Laboratory/California
Institute of Technology
Pasadena, California

Lucas T. Rich
Idaho National Laboratory
Idaho Falls, Idaho

Glenn R. Romanoski
Oak Ridge National Laboratory
Oak Ridge, Tennessee

Carl E. Sandifer
NASA Glenn Research Center
Cleveland, Ohio

Michael B.R. Smith
Oak Ridge National Laboratory
Oak Ridge, Tennessee

George B. Ulrich
Oak Ridge National Laboratory
Oak Ridge, Tennessee

Christofer E. Whiting
University of Dayton Research Institute
Dayton, Ohio

David Friedrich Woerner
Jet Propulsion Laboratory/California
Institute of Technology
Pasadena, California

Andrew J. Zillmer
Idaho National Laboratory
Idaho Falls, Idaho

Acknowledgments

I wish to acknowledge my parents, Robert Woerner and Mary Crow. It is their devotion to me, love, and upbringing that made my participation in this project possible.

I want to thank Brett Kurzman and Stacey Woods, our editors, and their colleagues at Wiley.com. They were responsive, dedicated, patient, and extremely helpful in preparing this manuscript.

I want to thank the small army of women and men that made this publication possible, those tireless souls that founded, grew, and built the Radioisotope Power System community into what it is today. The list of their names could fill the pages of this book, and I struggled with who to single out and mention here and concluded I would err on the side of not slighting anyone and just say thank you to the entire RPS community. Thank you. Thank you. You gave me friends, companions, and a career I did not expect.

I am deeply humbled and fascinated by the work and achievements of the RPS community and hope that this manuscript does them at least some small amount of justice. I know the authors have poured their hearts into the book and deserve a round of applause.

I/we are indebted to the reviewers whose clarity of mind and perceptive insights kept us focused and out of the "ditch" of lousy writing.

Lastly, I want to thank those space exploring scientists and engineers who thrill the world with their fabulous discoveries. The draft of the next Planetary Science and Astrobiology Decadal Survey was released less than a handful of days before I typed these words. The draft promises that the search for past and present life in our solar system will continue. It describes new and exciting scienic missions to come in the next decade should Congress and NASA fund them. Several missions to destinations in our solar system are only achievable by RPS powered robots. Expect grand findings.

A portion of this research was carried out at the Jet Propulsion Laboratory, California Institute of Technology, under a contract with the National Aeronautics and Space Administration (80NM0018D0004).

Portions of this work were authored under the support of Oak Ridge National Laboratory managed UT- Battelle, LLC, under contract DE-AC05-00OR22725 with the US Department of Energy (DOE).

Portions of this work were authored thanks to the generous support of the University of Dayton Research Institute.

A portion of this research was carried out at the Idaho National Laboratory, under a contract with the US Department of Energy (DE-AC07-05ID14517).

Glossary

^{90}**Sr**—An isotope of strontium in fuels used in legacy terrestrial radioisotope power systems. ^{90}Sr is responsible for generating most of the thermal energy found in SrTiO3 and SrF2 fuels used in radioisotope power systems.

^{238}**Pu**—The predominant isotope of plutonium in fuels for spaceflight radioisotope power systems. ^{238}Pu is responsible for generating most of the thermal energy found in plutonium dioxide fuels used in radioisotope power systems.

^{241}**Am**—An isotope of synthetic radioactive element americium. Americium is a transuranic member of the actinide series in the periodic table. ^{241}AmO2 is used in most household smoke detectors as an ionizing radiation source. ^{241}Am is a heat producing radioisotope that has a lower specific thermal power, but may be cheaper and easier to produce in the European community.

Absorbed dose—The amount of energy absorbed by a given material (per unit mass) from ionizing radiation. The SI unit for absorbed dose is the *gray* (Gy), where $1 \text{ Gy} = 1 \text{ J·kg}^{-1}$. Non-SI units for absorbed dose are *rad,* where $1 \text{ Gy} = 100$ rad.

Atomic Energy Commission (AEC)—An agency of the United States government established in 1946 for the development of atomic science and technology. The AEC was abolished in 1974, and its functions were divided between two new agencies: the Energy Research and Development Administration (ERDA) and the Nuclear Regulatory Commission (NRC).

Atomic notation—While complete notation of a given atom may include the atomic number Z, the mass number A, and the neutron number N (e.g., $^{A}X_{N}$), it is common practice to exclude Z and N and refer to a given isotope only by the elemental symbol and A, such as ^{238}Pu, ^{241}Am, or ^{90}Sr.

Beginning of Life (BOL)—The point in time at which an RPS is fully fueled.

Beginning of Mission (BOM)—The point in time at which a spacecraft has separated from its launch vehicle and the RTG(s) it carries is thermally stable. Note: BOM for Mars landers has traditionally been reported at the time of landing and not launch. This is the case for the Viking 1 and 2 landers, Curiosity, and Perseverance.

Cold cap—An electrically conductive cap bonded to a leg of a thermoelectric couple and used to bond to a cold strap.

Cold strap—An electrical connector which connects the cold caps of adjacent thermoelectric legs and couples in series and/or parallel to form the path for the electrical power generated.

Department of Energy (DOE)—A department in the United States government formed in 1977 when the Energy Research and Development Administration (ERDA), the Federal Energy Administration (FEA), the Federal Power Commission (FPC), and various other Federal agencies were reorganized.

Effective dose—A complex biological dose quantity representing the probability of a person to develop cancer from a given exposure to ionizing radiation. While the base units for *effective* dose are the same as *absorbed* dose ($J \cdot kg^{-1}$), effective dose considers additional weighting factors for radiation type and whole- body tissue types. The SI unit for effective dose is the *sievert* (Sv), where $1 \text{ Sv} = 1 \text{ J} \cdot kg^{-1}$. Non-SI units for absorbed dose are *rem,* where $1 \text{ Sv} = 100 \text{ rem}$.

End of Design Life (EODL)—The point in time marked by the end of the time span for which an RTG was designed. EODL is measured from the BOL of an RPS.

End of Mission (EOM)—The point in time at which a spacecraft completes its *primary* mission.

Dimensionless Figure of Merit (ZT)—The relative effectiveness of a thermoelectric material is the ratio of the product of the material's electrical conductivity (σ), square of the Seebeck coefficient (S), and temperature (T), to the thermal conductivity (κ). When discussing figure of merit, either z or Z can be used, though Z is more common. Mathematically the dimensionless Figure of Merit is represented as:

$$zT = \frac{\sigma S^2 T}{\kappa}$$

Fluence—A unit of measure to represent the cumulative number of particles that passed through a given surface area over a certain amount of time. Typically, fluence is expressed in units of *particles*cm^{-2}*, or *#·cm^{-2}*.

Flux—A unit of measure to represent the rate of particles passing through a given surface area. Typically, flux is expressed in units of $particles*cm^{-2} \cdot s^{-1}$, $\# \cdot cm^{-2} \cdot s^{-1}$.

Fuel clad—A material used to encapsulate a fuel pellet for an RPS. An alloy of iridium, DOP-27, is used to encapsulate fuel in a GPHS. The iridium cladding container must meet stringent safety requirements. This should not be confused with a fueled clad, which is a fuel pellet encapsulated in a cladding.

Fuel pellets—A pressed pellet of radioisotopic fuel.

General Purpose Heat Source (GPHS)—A radioactive heat source for radioisotope power systems (RPS). The heat source is comprised of four iridium alloy-clad plutonium-238 dioxide pellets. The encapsulated pellets are encased within nested layers of carbon-based materials and placed within an aeroshell assembly to comprise a complete GPHS. GPHSs can withstand extreme conditions including a launch-pad explosion or a high-speed re-entry.

General Purpose Heat Source RTG or GPHS-RTG—A radioisotope thermoelectric generator (RTG) design used on US space missions launched from 1989 thru 2006. It was fueled with eighteen GPHSs.

Glenn Research Center (GRC)—One of ten major NASA facilities, whose primary mission is to develop science and technology for use in aeronautics and space. Specific to radioisotope power systems, the center manages all of NASA's activities related to radioisotope power systems.

Half-life—The time it takes for an initial quantity of radioactive atoms to decrease in number (via radioactive decay) to half of the original quantity. All radioisotopes have a unique and well-defined half- life.

Heat rejection system—Those components which provide a means of dissipating the rejected heat from a thermoelectric converter.

Idaho National Laboratory (INL)—A federally funded research and development center (FFRDC) sponsored by the US Department of Energy (DOE). Specific to radioisotope power systems, the laboratory conducts activities related to the assembly and testing and delivery of radioisotope power systems.

Integrated ZT or Zint—The integrated thermoelectric figure of merit over the temperature gradient ($\Delta T = T_h - T_c$) across a thermoelectric converter:

$$Z_{int} = \left(\frac{1}{\Delta T} \right) \int_{T_c}^{T_h} Z(T) dT$$

Interagency Nuclear Safety Review Board (INSRB)—An intragovernmental advisory committee comprised of full-time officers or employees of the Federal Government. The INSRB conducted independent assessments of the safety analysis of space launches involving RPS and reactor units.

Isotope—All atoms with the same number of protons (Z) are considered the same *element*. However, atoms with the same number of protons but different numbers of neutrons (N) are called *isotopes* of the same *element*. For example, ^{236}Pu and ^{238}Pu are two different *isotopes* of the same *element*—Pu.

Jet Propulsion Laboratory (JPL)—A federally funded research and development center (FFRDC) and NASA field center. The laboratory's primary function is the construction and operation of planetary robotic spacecraft used to conduct scientific experiments. JPL has supported RTG development, testing, and operations for decades.

Johns Hopkins Applied Physics Laboratory (APL)—A not-for-profit university-affiliated research center (UARC) that has built and operated many spacecraft. APL was an early adopter of RTG technology and has been actively engaged in the development and use of RTGs.

Long-term testing—Testing which is carried out to determine and quantify performance characteristics and degradation mechanisms for thermoelectric materials, elements, couples, modules, and technologies over a long period of time. Data thus accumulated can be used to bolster confidence in performance predictions that extrapolate until the EODL.

Los Alamos National Laboratory (LANL)—A federally funded research and development center (FFRDC) sponsored by the US Department of Energy (DOE). Specific to radioisotope power systems, the laboratory conducts activities related to the processing and encapsulation of radioisotope fuel.

Mission Life (ML)—The period of time between launch of a spacecraft and the end of its primary mission.

Modular—Is the property of an RTG that allows it to be assembled into one or more variants. Each variant is fueled with a different number of GPHSs and relies upon the same converter technology.

Multi-Mission RTG (MMRTG)—A radioisotope thermoelectric generator (RTG) design used on US space missions launched from 2012 thru present day. It is fueled with eight GPHS modules and uses PbTe/TAGS thermoelectric materials.

Next Generation RTG (NGRTG)—A high-power, vacuum-rated RTG under development to enable future deep space missions. The NGRTG design builds upon the GPHS-RTG design that powered the Ulysses, Galileo, Cassini, and New Horizons missions.

n-leg or n-element—n-type thermoelectric material fabricated into one leg or element of a thermoelectric couple.

NSPM-20—A Presidential Memorandum on Launch of Spacecraft Containing Space Nuclear Systems. This memorandum updated the federal process by establishing a risk-based safety analysis and launch authorization processes.

Nuclear Power Assessment Study (NPAS)—A NASA study that identifies opportunities and challenges of a sustainable provisioning strategy for safe, reliable, and affordable nuclear power systems for the next 20 years.

Oak Ridge National Laboratory (ORNL)—A federally funded research and development center (FFRDC) sponsored by the US Department of Energy (DOE). Specific to radioisotope power systems, the laboratory conducts activities related to the production of radioisotope fuel, and fabrication of metal alloys and carbon materials.

p-leg or p-element—p-type thermoelectric material fabricated into one leg or element of a thermoelectric couple.

Planetary Science Decadal Survey—One in a series of studies funded by the NASA Science Mission Directorate and generated by the National Academies of Sciences, Engineering, and Medicine that identifies a research strategy to maximize advancement of planetary science, astrobiology, and planetary defense over a ten-year period. The survey categorizes and prioritizes scientific missions and goals.

Plutonium—A radioactive metallic element with the atomic number 94. It was discovered in 1940 by a team led by Glenn Seaborg.

Plutonium (di)oxide—Plutonium dioxide is a chemically stable ceramic material with an extremely low solubility in water and a high melting point ($2,744\,^\circ$C). Plutonium metal spontaneously oxidizes to $PuO2$ in an atmosphere containing oxygen.

Plutonium isotopes—There are five "common" isotopes of plutonium found in RTG fuel, ^{238}Pu, ^{239}Pu, ^{240}Pu, ^{241}Pu, and ^{242}Pu.

Radioisotope—Certain isotopes are stable and will remain in their original nuclear configuration indefinitely. However, certain isotopes are unstable—or *radioactive*—and referred to as *radioisotopes*.

Radioisotope Power Systems (RPSs)—Power generators that convert the energy available from radioisotopic fuels into electricity. RTGs are a type of RPS, but the term RPS also includes non- thermoelectric conversion technologies, including: heat engines (Sterling, Rankine, Brayton, etc.), thermionic, and others. To date, RTGs are the only RPS technology that has successfully flown in space.

Radioisotope Thermoelectric Generators (RTGs)—Power generators that convert the thermal energy available from radioisotopic fuels into electricity using thermoelectric converters.

Reactor or nuclear reactor—A reactor is a machine used to initiate and control a fission nuclear chain reaction.

RTG efficiency—Power output of an RTG divided by heat input.

Segmented leg—A thermoelectric leg formed of two or more different thermoelectric materials, bonded together.

Systems for Nuclear Auxiliary Power (SNAP)—A program for the development of RTGs and space nuclear reactors flown during the 1960s and 1970s.

Technology Readiness Level (TRL)—A systematic, metrics-based approach to assess the maturity of, and the risk associated with, a particular technology under development.

Thermal inventory—The total heat in an RTG. The total is maximum at the time an RTG is fueled. The thermal inventory of the fueled RTG decays from that time forward. This is approximately 0.8% per year for the plutonium fuel used in RTGs today.

Thermoelectric converter—A device that converts heat to electricity using multiple thermoelectric couples, which are arranged in arrays or modules.

Thermoelectric converter efficiency—Efficiency estimated for the thermoelectric converter is done using this equation:

$$\eta = \frac{\left(T_h - T_c\right)}{T_h} * \frac{\sqrt{1 + ZT} - 1}{\sqrt{1 + ZT} + \dfrac{T_c}{T_h}}$$

Where T_h and T_c are the hot- and cold-junction temperatures respectively, and ZT is the integrated dimensionless thermoelectric figure of merit. This efficiency estimate does not include system-level effects such as power losses in harnesses. This equation combines temperature gradients and ZT only.

Thermoelectric couple—A device that converts heat to electricity using the Seebeck effect. It is typically comprised of p- and n-legs electrically connected in series and thermally in parallel.

Thermoelectric couple hot-side interconnect—A connector between the hot junctions of p- and n-type thermoelectric legs in series to form an electrical path.

University of Dayton Research Institute (UDRI)—A national leader in scientific and engineering research. Specific to radioisotope power systems, the institute maintains hands-on RTG expertise that dates back to the Galileo/Ulysses missions. UDRI also conducts research and testing activities related to radioisotope fuel and encapsulation materials, thermoelectric materials and devices, and long-term RTG performance.

List of Acronyms and Abbreviations

3M	Minnesota Mining and Manufacturing Company
AEC	Atomic Energy Commision
ALSEP	Apollo Lunar Surface Experiments Package
APL	Applied Physics Laboratory
AR	Aerojet Rocketdyne
ARC	Ames Research Center
ATLO	Assembly, Test, and Launch Operations
ATR	Advanced Test Reactor
AU	Astronomical Unit
BOL	Beginning of Life
BOM	Beginning of Mission
CBCF	Carbon Bonded Carbon Fiber
CCAFS	Cape Canaveral Air Force Station
CFC	Chlorofluorocarbon
CFR	Code of Federal Regulations
CG	Center of Gravity
CLPS	Commercial Lunar Payload Services
CM	Configuration Manager
CMOS	Complementary Metal Oxide Semiconductor
COC	Certificate of Compliance
COPV	Composite Over-Wrapped Pressure Vessel
CSCBA	Converter Shipping Container Base Assembly
DARPA	Defense Advanced Research Projects Agency
DC	Direct Current
DoD	Department of Defense
DOE	Department of Energy
DOS	Density of States
DOT	Department of Transportation
DPA	Destructive Physical Analysis

DSA	Design Safety Analysis
ECR	Electrical Contact Resistance
EHS	Electric Heat Source
EM	Electromagnetic
EMI/EMC	Electromagnetic Interference/Electromagnetic Compatibility
eMMRTG	Enhanced Multi-Mission Radioisotope Thermoelectric Generator
ENDF	Evaluated Nuclear Data File
EODL	End of Design Life
EOM	End of Mission
ESA	European Space Agency
ESD	Electrostatic Discharge
ETG	Electrically Heated Thermoelectric Generator
EU	Engineering Unit
FAA	Federal Aviation Administration
FC	Fuel Clad
FET	Field Effect Transistor
FWPF^{TM}	Fine Weave Pierced^{TM} Fabric
GE	General Electric
GFY	Government Fiscal Year
GIS	Graphite Impact Shell
GLFC	Graphite LM Fuel Cask
GPHS	General Purpose Heat Source
GPHS-RTG	General Purpose Heat Source Radioisotope Thermoelectric Generator
GPS	Global Positioniong System
GRC	Glenn Research Center, NASA
GSE	Ground Support Equipment
GSFC	Goddard Spaceflight Center, NASA
GUI	Graphical User Interface
HBM	Human-Body Model
HDBK	Handbook
HEOMD	Human Exploration and Operations Mission Directorate
HEPA	High-Efficiency-Air-Particulate
HFIR	High Flux Isotope Reactor
HRS	Heat Rejection System
HS	Heat Source
I&T	Integration and Test
IAAC	Inert Atmosphere Assembly Chamber
ICV	Inner Containment Vessel
IECEC	International Energy Conversion Engineering Conference

INL	Idaho National Laboratory, DOE
INSRB	Interagency Nuclear Safety Review Board
IRD	Interface Requirements Document
IRHS	Intact Reentry Heat Source
ISPM	International Solar Polar Mission
ISRO	Indian Space Research Organization
JHUAPL	Johns Hopkins University Applied Physics Laboratory
JPL	Jet Propulsion Laboratory/California Institute of Technology
KAERI	Korea Atomic Energy Research Institute
KBO	Kuiper Belt Object
KSC	Kennedy Space Center
LANL	Los Alamos National Laboratory, DOE
LDS	Ling Dynamic System
LES	Lincoln Experimental Satellite
LET	Linear Energy Transfer
Li-ion	Lithium Ion
LMTO	Linear Muffin-Tin Orbital
LPPM	Life Performance Prediction Model
LSP	Launch Services Program
LTOF	Lift Turnover Fixture
LV	Launch Vehicle
LWRHU	Lightweight Radioisotope Heater Unit, aka RHU
MACS	Medium Altitude Communications Satellite
MHW	Multi-Hundred-Watt
MHW HS	Multi-Hundred-Watt Heat Source
MHW-RTG	Multi-Hundred-Watt-Radioisotope Thermoelectric Generator
MIL	Military
min	minimum; except when min is clearly being used to denote the time unit minute
MIT	Massachusetts Institute of Technology
MITG	Modular Isotopic Thermoelectric Generator
ML	Mission Life
MMAS	Martin Marietta Astro Space
MMLT	Mini-Modules Life Tester
MMRTG	Multi-Mission Radioisotope Thermoelectric Generator
MRM	Module Reduction and Monitoring
MSL	Mars Science Laboratory
NASA	National Aeronautics and Space Administration
NEPA	National Environmental Policy Act
NETS	Nuclear and Emerging Technologies for Space Conference
Next Gen RTG	Next Generation Radioisotope Thermoelectric Generator

NGRTG	Next Generation Radioisotope Thermoelectric Generator
NH	New Horizons
NNDC	National Nuclear Data Center
NNL	National Nuclear Laboratory
NOAA	National Oceanic and Atmospheric Administration
NPAS	Nuclear Power Assessment Study
NPS	Nuclear Power System
NRC	Nuclear Regulatory Commission
NSPM	National Security Presidential Memoranda
ORNL	Oak Ridge National Laboratory, DOE
PAWS	Powered Polar Automated Weather Station
PGS	Power Generation and Storage
PHSF	Payload Hazardous Servicing Facility
PMC	Plutonia Molybdenum Cermet
PMP	Portable Monitoring Package
PPO	Pure Plutonium Oxide
PRT	Platinum Resistance Thermometer
PSD	Planetary Science Division, NASA
QU	Qualification Unit
REDC	Radiochemical Engineering Development Center, ORNL
RFP	Request For Proposal
RHU	Radioisotope Heater Unit
RIC	RTG Integration Cart
RPS	Radioisotope Power System
RPS-DET	Radioisotope Power System Dose Estimation Tool
RSICC	Radiation Safety Information Computational Center
RTG	Radioisotope Thermoelectric Generator
RTGF	Radioisotope Thermoelectric Generator Facility
RTGTS	Radioisotope Thermoelectric Generator Transportation System
SAM	Sample Analysis Mars
SAR	Safety Analysis Report
SARC	Safety Analysis Report Commitment
SEB	Single-Event Burnout
SEE	Single-Event Effect
SEFI	Single-Event Functional Interupt
SEGR	Single-Event Gate Rupture
SER	Safety Evaluation Report
SET	Single-Event Transient
SEU	Single-Event Upset
SIG	Selenide Isotope Generator
SiGe	Silicon germanium

SKD	Skutterudite
SMD	Science Mission Directorate, NASA
SNAP	Systems for Nuclear Auxiliary Power
SNS	Space Nuclear System
SOW	Statement of Work
SPF	Single Point Failure
SRG	Stirling Radioisotope Generator
SRS	Savannah River Site
SSPSF	Space and Security Power Systems Facility
STD	Standard
SUV	Sports Utility Vehicle
SV	Space Vehicle
SwRI	Southwest Research Institute
TAGS	Tellurium antimony germanium silver alloy
TBR	To Be Reviewed
TCR	Thermal Contact Resistance
TE	Thermoelectric
TEC	Thermoelectric Couple
TEG	Thermoelectric Generator
TEM	Thermoelectric Multicouple
TESI	Teledyne Energy Systems, Incorporated
TRN	Terrain Relative Navigation
TSOC	Test and Operations Support Contract
TVAC	Thermal Atmosphere Vacuum Chamber
UDRI	University of Dayton Research Institute
UK	United Kingdom
US	United States
USAF	United States Air Force
USN	United States Navy
VA	Verification Activity
VCE	Voltage, Collector-Emitter
VDS	Volts, Drain-Source
VHP	Vaporous Hydrogen Peroxide
VIF	Vertical Integration Facility
wrt	with respect to

1

The History of the Invention of Radioisotope Thermoelectric Generators (RTGs) for Space Exploration

Chadwick D. Barklay

University of Dayton Research Institute, Dayton, Ohio

In December of 1903, the Wright Brothers made the first successful powered flight of an airplane. There are significant levels of examination of Orville and Wilbur's incremental improvements to the original design of their flying machine to build the Wright Flyer II. However, there is not much appreciation of the backstory that inspired the brothers to explore the fundamentals of aerodynamics and pursue the research and development required to make a powered, heavier-than-air aircraft. Wilbur Wright indicated in a letter he wrote in 1912 that the pioneering work of Otto Lilienthal in the late 1800s was a precursor to their efforts. But it was a rubber band-powered toy helicopter their father, Milton Wright, gave them in 1878 that Orville credited as the object that sparked their interest in flight.

As an opening discussion of the history of the radioisotope thermoelectric generator (RTG), it is essential to understand the backstory of the invention that has allowed humankind to explore beyond the solar system's boundaries. In 1954, Kenneth Jordan and John Birden invented the RTG at the Atomic Energy Commission (AEC) Mound Laboratory. Oral history posits that the two inventors drafted their initial design concept during lunch on a napkin in the Mound Laboratory cafeteria. Their initial research efforts used a small steam-electric generator to demonstrate that heat utilized from the radioactive decay of polonium-210 could generate electricity. However, more-efficient methods for producing electricity were required, and Jordan and Birden coupled a polonium-210 heat source to a thermoelectric material array to generate electricity (Figure 1.1). This early prototype used forty chromel-constantan thermocouples to generate power from a suspended sphere containing 146 curies of encapsulated polonium-210. The outside container of the prototype was made of aluminum and used an early

The Technology of Discovery: Radioisotope Thermoelectric Generators and Thermoelectric Technologies for Space Exploration, First Edition. Edited by David Friedrich Woerner.

Figure 1.1   An early experimental thermoelectric generator designed by Jordan and Birden that couples a polonium-210 heat source with a thermoelectric material array. Credit: Mound Science and Energy Museum Association.

form of silica aerogel (Santocel) as insulation. This unit produced 9.4 milliwatts of power for a total efficiency of 0.20%. [1] In 1959, Jordan and Birden received a patent for their invention, which is still the underpinning innovation for all RTGs used by the National Aeronautics and Space Administration (NASA) for planetary and deep-space exploration. [2]

Similarly, the events leading up to the first powered flight at Kittyhawk, the backdrop of Jordan and Birden's early efforts, are not widely known. The US Congress established the AEC in the shadows of World War II to establish centralized governmental controls to manage the research and production of atomic weapons in the post-war era. [3] During the first decade of the AEC, laboratories under the AEC umbrella conducted a broad spectrum of research activities on producing natural and synthetic radioisotopes in reactors and cyclotrons. These early efforts focused on how radioisotopes can influence thermonuclear fusion produced by weapons. In conjunction with research and development activities for weapons programs, research was also ongoing to determine the utility of radioisotopes for particle physics, medicine, geography, and several industrial applications.

Within the same timeframe, the War Department, the predecessor to the Department of Defense, recognized a need for establishing a non-profit, global policy think tank, and the Douglas Aircraft Company created the RAND Project, later the RAND Corporation, to fulfill this need. In 1946, the RAND Project explored the preliminary design of a satellite vehicle [4], and in 1947, RAND expanded its examination to evaluate the use of radioisotopes to address the electrical power requirements for satellite vehicles. [5] This initial analysis considered the use of polonium-210 and strontium-89 as thermal sources for power generation. In 1949, RAND published a study that outlined the use of nuclear power sources for satellites in Earth orbit, and in 1951, the Department of Defense (DOD) requested that the AEC initiate research on nuclear power for spacecraft. [6] As a result, the AEC initiated a series of studies that concluded that both fission and radioisotope power systems were technically feasible for satellites. [7]

In late 1953, President Dwight D. Eisenhower delivered his "Atoms-for-Peace" address at the United Nations to promote the peaceful uses of atomic energy. During that same timeframe, Jordan and Birden built two experimental thermal batteries using polonium-210 and chromel-constantan thermocouples to validate their thermal battery theory and develop fabrication techniques. These experimental units had approximately ten times the work capacity of ordinary dry cell batteries of the same weight. [1]

Jordan and Birden fabricated seven experimental units in total, and the third unit (Figure 1.2) was the prototype of the remaining generators built. These later

Figure 1.2 One of the latter prototype thermoelectric generators designed by Jordan and Birden. Credit: Mound Science and Energy Museum Association.

units employed twelve thermopiles, each consisting of thirty-seven chromel-constantan thermocouples supported by mica cards that were vertically mounted and radially spaced on aluminum rings. Figure 1.2 also shows an assembled and disassembled thermopile.

Jordan and Birden made extensive measurements on these units to determine efficiencies and the effects of various types of insulation, including vacuum, noise levels, ambient temperatures, matched and unmatched loads on the units' performance. [8] In general, these latter prototypes produced approximately 50 milliwatts of power for a total efficiency of 0.32%. The last units in this series of prototypes were designed and built based on specifications that delineated load voltage, power, durability, and design life requirements. It was the first step to ensuring that future generators would be sufficiently rugged to withstand the vibrational and quasi-static forces associated with space launches.

In 1955 the AEC formally initiated the Systems for Nuclear Auxiliary Power (SNAP) program, which focused on experimental radioisotope and fission systems. The objective of this program was to develop compact, lightweight, reliable atomic electric devices for use in space and terrestrial applications. Under the SNAP program, odd numbers designated RTGs systems, and even numbers represented fission reactor systems. In 1957, the Martin Company developed the SNAP-1 RTG, which was assembled at the AEC Mound Laboratory. As the development of SNAP-1 progressed, the Martin Company subcontracted with Westinghouse Electric and the Minnesota Mining and Manufacturing Company (3M) to develop the SNAP-3 RTG. In 1958, 3M delivered the SNAP-3 to the Martin Company, which fueled the unit with encapsulated polonium-210 from Mound Laboratory. The SNAP-3 produced 2.5 W_e, and President Eisenhower displayed the power system in the US White House's Oval Office on 16 January 1959 (Figure 1.3). [9]

In 1961, the US Air Force launched the Transit 4A and 4B naval navigation satellites from Cape Canaveral Air Force Station. The Navy Transit program required a power source that would reliably operate for five years. To meet this requirement, the AEC modified a SNAP-3 to utilize plutonium-238 rather than polonium-210. This modification extended the design life of the SNAP-3A because plutonium-238 has a longer half-life than polonium-210 (88 years versus 138 days for polonium-210). [10] The Transit 4A and 4B satellites utilized a hybrid power system comprised of a SNAP-3B RTG and solar panels, which provided a total power of 35 W_e. [10] These early navigation satellites were a precursor to the global positioning system (GPS), which provides positioning and navigational capabilities to military, civil, and commercial users worldwide. In addition, the successful mission of these satellites demonstrated the feasibility of using RTGs for space missions.

Figure 1.3 The public debut of the SNAP-3 RTG technology demonstration device in the US White House's Oval Office on 16 January 1959. Pictured left to right: President Eisenhower, Major General Donald Keirn, AEC Chairman John McCone, Colonel Jack Armstrong, and Lt. Colonel Guveren Anderson. Credit: DOE Flickr.

All modern aircraft incorporate some of the essential design elements of the 1903 Wright Flyer, and correspondingly all RTG designs have their antecedents in a thermal battery concept on a cafeteria napkin. The legacy of Jordan and Birden's creative efforts transcend deep space and planetary exploration and has an ongoing impact on technology, culture, art, and literature.

References

1 Jordan, K.C. and Birden, J.H. (1954). Thermal Batteries Using Polonium-210. *MLM-984*. Mound Laboratory, Miamisburg, OH. https://doi.org/10.2172/4406897
2 Birden, J.H. and Jordan, K.C. (1959). Radioactive battery. US Patent 29, 135, 10A.
3 Federal register. (1947). *Federal Register*, vol. 12(2). Office of the Federal Register, National Archives and Records Administration, https://www.govinfo.gov/content/pkg/FR-1947-11-18/pdf/FR-1947-11-18.pdf (accessed 29 November 2022).

4 Clauser, F.H. et al. (1946). Preliminary Design of an Experimental World-Circling Spaceship. *SM-11827*, p. 340. Douglas Aircraft Company, Santa Monica, CA. https://www.rand.org/pubs/special_memoranda/SM11827.html (accessed 29 November 2022).

5 Avalrez, L.W. et al. (1947). Use of Nuclear Energy in the Satellite. *RA-15032*, p. 60. Douglas Aircraft Company, Santa Monica, CA. https://www.governmentattic.org/RAND/RA-15032.pdf (accessed 29 November 2022).

6 Dick, S.J. (ed.) (2015). *Historical Studies in the Societal Impact of Spaceflight*. Washington, DC: National Aeronautics and Space Administration Office of Communications, NASA History Program Office.

7 Corliss, W.R. (1966). *SNAP Nuclear Space Reactors*, 51. Washington, DC: Atomic Energy Commission https://www.osti.gov/servlets/purl/1132526.

8 Blanke, B.C., Birden, J.H., Jordan, K.C., and Murphy, E.L. (1960). Nuclear Battery-Thermocouple Type Summary Report. *MLM-1127*. Mound Laboratory, Miamisburg, OH. https://doi.org/10.2172/4807049.

9 Engler, R. (1987). *Atomic Power in Space: A History*. DOE/NE/32117-H1. Washington, DC: US Department of Energy.

10 Morse, J.G. (1963). Radioisotope-Fueled Power Supplies. *TID-18323*. Martin Company, Baltimore, MD. https://doi.org/10.2172/4718080

2

The History of the United States's Flight and Terrestrial RTGs

Andrew J. Zillmer

Idaho National Laboratory, Idaho Falls, Idaho

Space nuclear power systems have their origins in works by Arthur C. Clarke and studies for surveillance and navigation satellites produced after World War II. The processing of materials for the building of atomic bombs led to a number of radioisotopes being available, which in turn increased the number of radioisotopes available for alternative civilian and defense purposes. [1] The Systems for Nuclear Auxiliary Power (SNAP) program consolidated several of the Atomic Energy Commission's (AEC) reactor and radioisotope programs in the mid-1950s. RPS models were identified by the odd numbers in their designations and reactor systems with even numbers.

2.1 Flight RTGS

2.1.1 SNAP Flight Program

Initial requirements for SNAP systems came from Project Feedback, a Rand Corporation study, that originated in 1946 to evaluate the feasibility of artificial satellites. Initial power requirements were at least 1 kW$_e$ of power with a closed cycle power plant operating between 422 °C (900 °F) and 704 °C (1300 °F). The original SNAP-1 and 2 systems used similar mercury Rankine dynamic power conversion technologies, and the SNAP-1 series of power systems included RTGs. [2]

2.1.1.1 SNAP-3

The SNAP-3 system was the first Radioisotope Thermoelectric Generator (RTG) deployed under the SNAP program. It used a static thermoelectric converter developed by 3M. The Martin Company Nuclear Division took delivery of the generator in late 1958 and fueled it with a capsule of polonium-210. The SNAP-3 RTG produced 2.5 watts of electricity. [3] Figure 2.1 shows a cross section of the SNAP-3 system.

The design was revised to use a plutonium-238 heat source. This new system was designated SNAP-3B. The heat source for the SNAP-3B was plutonium metal and operated at 537 °C (998 °F). The generator was designed to burnup during a hypersonic re-entry accident and disperse its fuel at high altitude in line with the aerospace nuclear safety approach of the day. In June 1961, the Transit 4A navigation satellite launched and carried a SNAP-3B power source. This was the first use of nuclear power in space. [4,5]

2.1.1.2 SNAP-9

The SNAP-9 RTG was an expanded design of the SNAP-3B and produced 25 watts of electricity and is in Figure 2.2. Fins were included in the SNAP-9 design as radiators, unlike the SNAP-3 series of RTGs, and the system was not hermetically sealed like the SNAP-3. The design life of the system was for a one year shelf life and a five year operational life. It used a plutonium-238 heat source that was

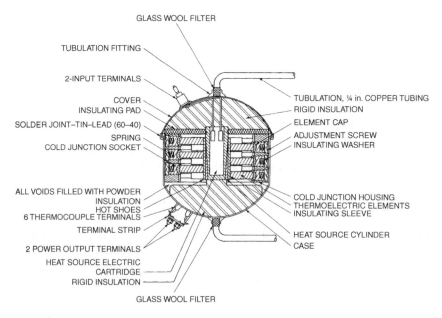

Figure 2.1 Cross section of a SNAP-3 RTG. [5] Credit: NASA.

Figure 2.2 A fully assembled SNAP-9 RTG. [6], Bennett 2006, American Institute of Aeronautics and Astronautics.

melted so that the fuel bonded to the fuel capsule liner. Bonding the heat source to the liner gave excellent heat transfer characteristics. SNAP-9A powered two successfully launched Transit 5BN navigation satellites and one Transit 5BN satellite that failed to achieve orbit. [3,6]

2.1.1.3 SNAP-19

The SNAP Program developed several designs for the 19-series. The SNAP-19B RTG was the first RTG launched on a NASA spacecraft. Development from concept to launch took almost 5 years of collaboration between the AEC, NASA, and industry partners. SNAP-19B also helped to better establish a process for the use of RTGs on missions, as it was the first RTG to:

- develop interface compatibility between the power source and spacecraft in the RTG design rather than previous spacecraft interfaces
- have a complete set of user specified requirements and have stakeholder input during development

The SNAP-19 RTGs were also the first operational RTG power systems to be designed with an intact re-entry design. This resulted from a change in the design philosophy of dispersing the radioisotope heat source to ensuring that the radioisotope heat source would remain intact during a launch accident re-entry and impact events. [7] The new heat source form, called the Intact Impact Heat Source, generated 645 watts of thermal power from the decay of plutonium-238. The SNAP-19B used PbTe and tellurides of antimony, germanium, and silver (Te-Ag-Ge-Sb) or PbTe-TAGS thermoelectric modules for power conversion. [4,7]

The first SNAP-19B flew on the Nimbus III meteorological satellite in April 1969. The design life of the system was to operate for 10,000 hours. It ran for over 20,000 hours. After ~20,000 hours, there was a sharp drop in power that was attributed to the thermoelectric couples sublimating and the failure of hot junction bonds because of an excessive leak of the internal cover gas. The design of the SNAP-19 was modified to correct these issues and was later used on the Pioneer 10 and 11 spacecraft and the Viking 1 and 2 spacecraft. [4] Cross-sectional views of a SNAP-19 and modified SNAP-19 are depicted in Figures 2.3 and 2.4.

The SNAP-19 RTGs used on the Pioneer 10 and 11 missions had a design life of 6 years and produced a power output of 41.2 watts of electricity when fueled. The SNAP-19 design was changed for the Viking 1 and 2 Martian landers to survive sterilization of the landers to meet planetary protection requirements, minimize power losses during Mars transit, and to operate under diurnal temperature cycles on the Martian surface. These units had a two-year design life and generated 42.5 watts of electricity once fueled. [4]

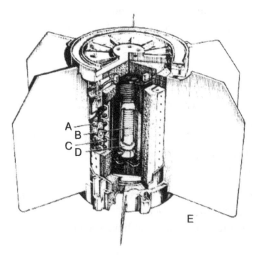

Figure 2.3 Cross section of an assembled SNAP-19 RTG as flown on the Pioneer interplanetary spacecraft. [8] Credit: NASA.

A THERMOELECTRICS
B FUEL CAPSULE
C REENTRY HEAT SHIELD
D FUEL DISCS
E HEAT RADIATING FINS

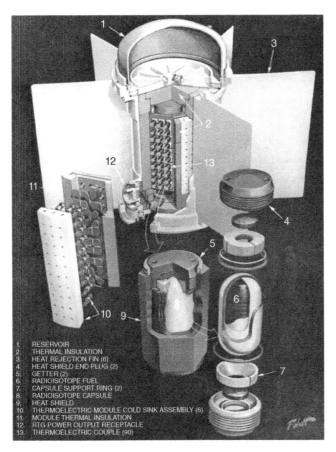

1. RESERVOIR
2. THERMAL INSULATION
3. HEAT REJECTION FIN (6)
4. HEAT SHIELD END PLUG (2)
5. GETTER (2)
6. RADIOISOTOPE FUEL
7. CAPSULE SUPPORT RING (2)
8. RADIOISOTOPE CAPSULE
9. HEAT SHIELD
10. THERMOELECTRIC MODULE COLD SINK ASSEMBLY (6)
11. MODULE THERMAL INSULATION
12. RTG POWER OUTPUT RECEPTACLE
13. THERMOELECTRIC COUPLE (90)

Figure 2.4 Major components of the modified SNAP-19 RTGs used on the Viking mission landers. Credit: Teledyne Isotopes.

2.1.1.4 SNAP-27

The SNAP-27 power system powered instrumentation and experiments on the lunar surface for the Apollo Program. It is the only RTG that astronauts have fueled. Figure 2.5 is a photograph of one being fueled on the surface of the moon. The SNAP-27 design minimized radiation exposure to the public in the event of a variety of accident scenarios that included ground impact, Earth atmosphere re-entry, launch abort, and explosions on the launch pad. [3,4] Figure 2.6 shows the major components of the SNAP-27.

Figure 2.5 Astronaut Alan Bean removing fuel from the lunar lander to insert into the SNAP-27 sitting just in front of him on the Moon's surface. Credit: NASA / Wikimedia Commons / CC0 1.0.

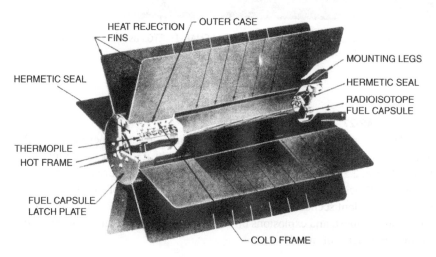

Figure 2.6 Diagram of the SNAP-27 RTG. Credit: US Department of Energy.

2.1.2 Transit-RTG

The Transit-RTG was developed for a navigational satellite and only flown once. The design goals for the Transit-RTG were a five-year life with a beginning of life power of 34.3 watts at 5 volts and weigh less than 17 kg. The thermoelectric converters used PbTe materials without a cover gas and were radiatively coupled to the heat source. The Transit-RTG used a heat source composed of a plutonium oxide molybdenum cermet that generated 850 watts-thermal. [4]

2.1.3 Multi-Hundred-Watt RTG

The Multi-Hundred-Watt (MHW) RTG operated with hotter heat sources compared to SNAP RTGs and silicon germanium (SiGe) thermoelectric material to reach an efficiency of 6.6%. Each MHW-RTG produced 158 watts of electricity once fueled. Figure 2.7 shows an assembled MHW-RTG.

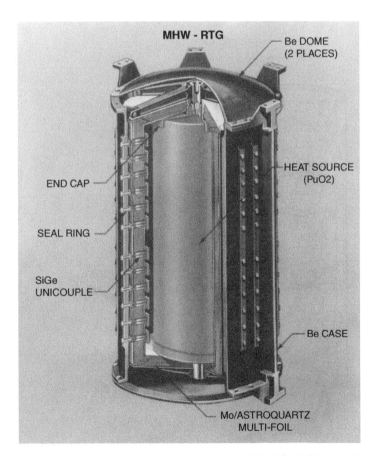

Figure 2.7 An assembled MHW RTG in cutaway. Credit: NASA / Wikimedia Commons / CC0 1.0.

Aerospace nuclear safety guidelines at the time drove the design of the heat source to immobilize plutonium in the heat source in nominal situations and accidents. The heat source design tended to prevent the release of the plutonium fuel, with a focus on minimizing the amount of biologically respirable particles. It would also minimize dispersion of the heat source materials and maximize the long term immobilization of the plutonium fuel in potential accident scenarios that might cause damage to the heat source. Incorporating these requirements into the design led to fuel sphere containers 3.7 cm in diameter that generated 100 watts of thermal power per sphere. Figure 2.8 shows a cutaway view of the MHW-RTG heat source assembly.

The MHW-RTG flew first on the Lincoln Experimental Satellites 8 and 9 (LES-8 and LES-9) and then the Voyager 1 and 2 spacecraft. The LES-8 and LES-9 satellites launched in March 1978, with LES-8 operating until 2004 and LES-9 until May 2020. Voyager 1 and 2 launched in the summer of 1977 and both are still operating as of July 2021. [3,4,10]

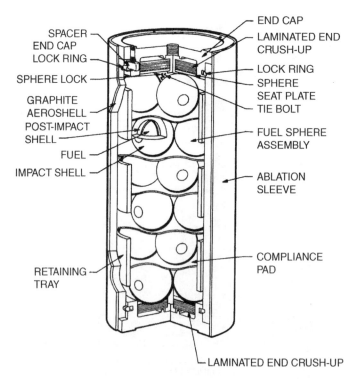

Figure 2.8 Cutaway view of the MHW-RTG heat source assembly. Credit: NASA.

2.1.4 General Purpose Heat Source RTG

The General Purpose Heat Source RTG (GPHS-RTG) was designed around a new heat source, the GPHS. The General Purpose Heat Source (GPHS) provided a standardized fuel form for RPS systems and improved the safety of the system. This contributed to the design of the GPHS-RTG with a higher specific power than an MHW-RTG or any earlier RTG and improved safety.

2.1.4.1 General Purpose Heat Source

The General Purpose Heat Source (GPHS) design added safety features not found in the earlier generation of spherical heat sources in the MHW-RTG. Figure 2.9 illustrates the major components of a GPHS. Each GPHS module produces approximately 250 watts of thermal power at its beginning of life. Each module is made from many components, including an outer aeroshell made from Fine Weave Pierced Fabric (FWPF) graphite to protect the fuel during a hypersonic re-entry and other potentially deleterious events, including thermal shock, heating, and ablation. Inside the aeroshell are two FWPF Graphite Impact Shells (GIS) that provide additional impact protection. Inside each GIS is a carbon bonded carbon fiber (CBCF) sleeve, which provides additional insulation. Each sleeve contains two plutonium-238 oxide fuel pellets. These pellets are clad in iridium alloy to provide containment impacts. [11]

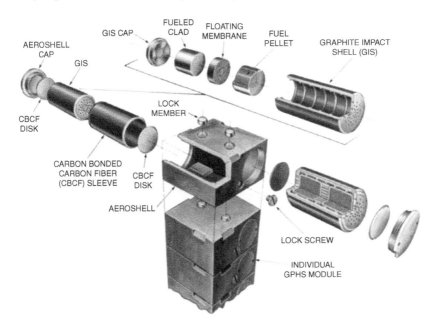

Figure 2.9 Expanded view of a GPHS module. Credit: DOE.

The GPHS module has evolved over its lifetime. The STEP 0 GPHS module is the original design, and was used on the Ulysses, Cassini, and Galileo missions. Each STEP 0 GPHS module weighs ~1.43 kg (3.15 lb). The STEP 1 GPHS module increases the width of the module and adds a web between the GIS sleeves to provide additional structural and thermal protection in accident scenarios. The STEP 1 GPHS module was used in the Pluto New Horizons mission and each module weighs ~1.51 kg (3.33 lb). The STEP 2 GPHS module keeps the web between the GIS sleeves, and increases the height of the module, which provides more protection in accident scenarios. The STEP 2 GPHS module weights ~1.61 kg (3.55 lb) when fueled. [12]

2.1.4.2 GPHS-RTG System

Each GPHS-RTG used 18 GPHS modules, producing approximately 4400 watts of heat in total to produce approximately 285 W of electric power when fueled. The thermoelectric couples in a GPHS-RTG are made of silicon germanium and use the same unicouple design as the MHW-RTG. None of the flown GPHS-RTGs were operated in a planetary atmosphere, such as on Mars, because the thermoelectric couples were not compatible with most atmospheres. As a result, the GPHS-RTG systems were used only in the vacuum of deep space. Figure 2.10 shows several major components of a GPHS-RTG.

GPHS-RTGs have flown on four missions. The Galileo mission used two GPHS-RTGs and launched from the Space Shuttle Atlantis in October 1989, and arrived at Jupiter in December 1995. Ulysses used one GPHS-RTG and performed solar observations. It launched in October 1990 and operated until June 2009. The Cassini mission launched in October 1997 and used three GPHS-RTGs. It arrived at Saturn in July 2004 and NASA sent it hurtling into Saturn's atmosphere in September 2017 to

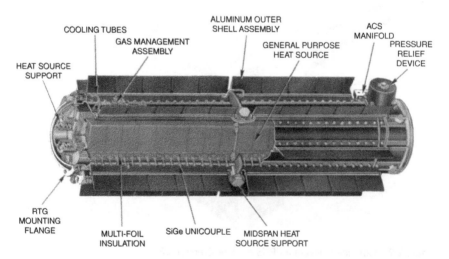

Figure 2.10 Diagram illustrating several major components of the GPHS-RTG. Credit: DOE.

protect the moons of Saturn from potential biological contamination. The New Horizons mission launched in January 2006 and used a single GPHS-RTG. New Horizons performed a flyby of Pluto and its moon Charon in July 2015 and encountered the Kuiper Belt object 486958 Arrokoth in January 2019. [3,13]

2.1.5 Multi-Mission Radioisotope Thermoelectric Generator

The Multi-Mission Radioisotope Thermoelectric Generator (MMRTG) has its origins in the early 2000s in a NASA RPS Provisioning Study, which helped to set the NASA and DOE strategy for future RTG development and identify science mission needs. The GPHS-RTG was designed to operate in vacuum, and not in planetary atmospheres, and generated approximately 300 watts of electricity when fueled. Forecasts about the power demand of future missions pointed to lower power needs and also indicated RTGs would be needed to operate in vacuum and in planetary atmospheres. The forecasts proved prescient. Designing a new RTG was also considered risk mitigation against a Stirling Radioisotope Generator (SRG), which was in development. [3,14]

In 2003, the DOE awarded a contract to the Rocketdyne division of Boeing to build the MMRTG. Their subcontractor Teledyne Energy Systems was to produce PbTe-TAGS thermoelectric couples similar to those that were used in the SNAP-19 RTGs. The final MMRTG design yielded a system that produces 110 watts of electricity from 8 GPHS modules. Figure 2.11 shows an MMRTG after fueling at Idaho National Laboratory.

Figure 2.11 The white MMRTG for the Perseverance Rover after fueling and sitting on the base of a transportation container. [16] Credit: [16] Cull (1989) US Department of Energy.

The MMRTG was first used on the Curiosity rover, which is part of the Mars Science Laboratory (MSL) mission. MSL launched from Cape Canaveral in November 2011 and landed on Mars in August 2012. A second MMRTG flew on the Perseverance rover, which is part of the Mars 2020 mission. Mars 2020 launched in July 2020 and landed in February 2021. Both the Curiosity and Perseverance rover, and their MMRTGs, have been in continual operation since they were fueled and they continue to operate on Mars as of the end of 2021. [15,16]

2.1.6 US Flight RTGs

The US has flown RTGs for more than six decades. A variety of designs have been used in space. That variety can be seen in Table 2.1 with one exception. The power conversion materials have remained largely constant since the first flight of an RTG, either PbTe or SiGe material systems have been used. No new material systems have been proven as reliable, but that may change in the near-future. Some of these systems are reviewed in Chapter 10.

2.2 Unflown Flight RTGs

2.2.1.1 SNAP-1
The Martin Company's Nuclear Division won the contract to build the SNAP-1 radioisotope power system in 1956. SNAP-1 was a ground demonstration only. The project began in 1957 and originally used a mercury Rankine power conversion cycle that was heated by the radioactive decay of cerium-144. The design goal of this system was to produce 400 watts of electricity throughout a lifetime of 60 days. The project was cancelled in 1959. [3]

2.2.1.2 SNAP-11
The SNAP-11 was developed as a power source for the lunar Surveyor missions. The RTG would power the landers throughout the lunar nights. The design used a curium-242 heat source to generate 25 watts of electricity with a design life of one quarter of a year. [17] They were never used on NASA's Surveyor missions.

2.2.1.3 SNAP-13
The SNAP-13 development was an evaluation of an evaluation an auxiliary space power source. The SNAP-13 concept used thermionic conversion technology to produce electricity from the decay of curium-242. The system produced 17 watts of power (24 amps at 0.5 volts) when it was fueled in November 1965. The generator failed after just 45 hours, well short of its design life of one quarter of a year. The likely cause of the failure was that the molten fuel leaked through its

Table 2.1 A catalog and summarization of general characteristics of US built RTGs that have flown.

Generator Type	Power Conversion Materials	Fuel Form	Power (We)[1] Nominal per RTG	Actual BOM Total	Conversion Efficiency[2]	Total Fuel Activity[3]	System Mass	Specific Power[4] (We/kg)	Spacecraft
SNAP-3B	2n/2p PbTe	^{238}Pu Metal	2.7	2.6	5.1%	6.66×10^{13} Bq	2.1 kg (4.6 lb)	1.3	Transit 4-A
SNAP-3B	2n/2p PbTe	^{238}Pu Metal	2.7	3.1	5.1%	6.66×10^{13} Bq	2.1 kg (4.6 lb)	1.3	Transit 4-B
SNAP-9A	2n/2p PbTe	^{238}Pu Metal	26.8	25.2	5.1%	5.92×10^{14} Bq	12.2 kg (27 lb)	2.2	Transit 5BN-1
SNAP-9A	2n/2p PbTe	^{238}Pu Metal	26.8	26.8	5.1%	5.92×10^{14} Bq	12.2 kg (27 lb)	2.2	Transit 5BN-2
SNAP-9A	2n/2p PbTe	^{238}Pu Metal	26.8	—[5]	5.1%	5.92×10^{14} Bq	12.2 kg (27 lb)	2.2	Transit 5BN-3
SNAP-19B	2n/3p PbTe	^{238}Pu Oxide Microspheres	27.5	—[5]	4.5%	1.26×10^{15} Bq	13.6 kg (30 lb)	2.1	Nimbus-B-1
SNAP-19B	2n/3p PbTe	^{238}Pu Oxide Microspheres	27.5	49.2	4.5%	1.39×10^{15} Bq	13.6 kg (30 lb)	2.1	Nimbus III
SNAP-27	3n/3p PbTe	^{238}Pu Oxide Microspheres	63.5	73.6	5.0%	1.65×10^{15} Bq	30.8 kg (68 lb)	2.3	Apollo 12
SNAP-27	3n/3p PbTe	^{238}Pu Oxide	63.5	—[5]	5.0%	1.65×10^{15} Bq	30.8 kg (68 lb)	2.3	Apollo 13

(Continued)

Table 2.1 (Continued)

Generator Type	Power Conversion Materials	Fuel Form	Power (W$_e$)[1]		Conversion Efficiency[2]	Total Fuel Activity[3]	System Mass	Specific Power[4] (W$_e$/kg)	Spacecraft
			Nominal per RTG	Actual BOM Total					
SNAP-27	3n/3p PbTe	Microspheres ^{238}Pu Oxide	63.5	72.5	5.0%	1.65×10^{15} Bq	30.8 kg (68 lb)	2.3	Apollo 14
SNAP-27	3n/3p PbTe	Microspheres ^{238}Pu Oxide	63.5	74.7	5.0%	1.65×10^{15} Bq	30.8 kg (68 lb)	2.3	Apollo 15
SNAP-19	2n PbTe/ TAGS-85	Microspheres ^{238}Pu PMC	40.3	162.8	6.2%	2.87×10^{15} Bq	13.6 kg (30 lb)	3.0	Pioneer 10
SNAP-27	3n/3p PbTe	^{238}Pu Oxide	63.5	70.9	5.0%	1.65×10^{15} Bq	30.8 kg (68 lb)	2.3	Apollo 16
Transit-RTG	2n/3p PbTe	Microspheres ^{238}Pu PMC	37.0	35.6	4.2%	9.48×10^{14} Bq	12.2 kg (27 lb)	2.6	TRIAD
SNAP-27	3n/3p PbTe	^{238}Pu Oxide Microspheres	69.0	75.4	5.0%	1.65×10^{15} Bq	30.8 kg (68 lb)	2.3	Apollo 17
SNAP-19	2n PbTe/ TAGS-85	^{238}Pu PMC	40.3	159.6	6.2%	2.87×10^{15} Bq	13.6 kg (30 lb)	3.0	Pioneer 11
SNAP-19	2n PbTe/ TAGS-85	^{238}Pu PMC	42.7	84.6[7]	6.2%	1.50×10^{15} Bq	13.6 kg (30 lb)	2.8	Viking Lander 1
SNAP-19	2n PbTe/ TAGS-85	^{238}Pu PMC	42.7	86.2[7]	6.2%	1.50×10^{15} Bq	13.6 kg (30 lb)	2.8	Viking Lander 2

RTG	Thermoelectric	Fuel				Activity	Mass		Mission
MHW-RTG	Si-Ge	^{238}Pu PPO	158	307	6.6%	4.90×10^{15} Bq	37.7 kg (83 lb)	4.2	LES 8
MHW-RTG	Si-Ge	^{238}Pu PPO	158	308	6.6%	4.90×10^{15} Bq	37.7 kg (83 lb)	4.2	LES 9
MHW-RTG	Si-Ge	^{238}Pu PPO	158	470	6.6%	7.36×10^{15} Bq	37.7 kg (83 lb)	4.2	Voyager 1
MHW-RTG	Si-Ge	^{238}Pu PPO	158	478	6.6%	7.36×10^{15} Bq	37.7 kg (83 lb)	4.2	Voyager 2
GPHS-RTG	Si-Ge	^{238}Pu PPO	292	576	6.6%	9.83×10^{15} Bq[6]	56 kg (123 lb)	5.1	Galileo
GPHS-RTG	Si-Ge	^{238}Pu PPO	292	289	6.6%	4.92×10^{15} Bq[6]	56 kg (123 lb)	5.0	Ulysses
GPHS-RTG	Si-Ge	^{238}Pu PPO	292	887	6.6%	1.47×10^{16} Bq	56 kg (123 lb)	5.3	Cassini
GPHS-RTG	Si-Ge	^{238}Pu PPO	237	246	6.0%	4.40×10^{15} Bq	56 kg (123 lb)	4.2	New Horizons
MMRTG	PbTe/TAGS-85	^{238}Pu PPO	110	115[7]	6.2%	2.23×10^{15} Bq	45 kg (99 lb)	2.8	Curiosity Rover
MMRTG	PbTe/TAGS-85	^{238}Pu PPO	110	118[7]	6.2%	2.23×10^{15} Bq	45 kg (99 lb)	2.8	Perseverance Rover

[1] Nominal power is based on either the design specification or the expected average power per RTG. Actual BOM is total power at beginning of mission.

[2] Based on nominal thermal wattage from the fuel and nominal RTG Power at BOM.

[3] Pure plutonium-238 produces 6.33×10^{11} Bq per gram of isotope. Plutonium-238 also produces 1 thermal watt per 1.11×10^{12} Bq of activity.

[4] These values are averages for BOM and may not represent the specification power for the RTG noted in column 1.

[5] Mission failed before RTG could be put in service.

[6] Activity at the point the RTGs were loaded with fuel. Does not account for the radioactive decay that occurred during the 5–6 year launch delay for these missions.

[7] Actual BOM power here is at landing on Mars; this tradition began with the Viking landers and is carried into the Curiosity and Perseverance landings; all other BOM power numbers are actuals measured after spacecraft launch and separation.

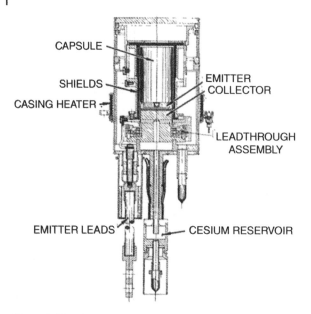

CAPSULE

SHIELDS

CASING HEATER

EMITTER
COLLECTOR

LEADTHROUGH
ASSEMBLY

EMITTER LEADS

CESIUM RESERVOIR

Figure 2.12 Cross-section of a SNAP-13 thermionic generator. [18] Credit: AEC/Martin Marietta.

encapsulation and damaged the diode spacer in the cesium envelope. A cross section of the SNAP-13 RTG is shown in Figure 2.12. [17, 18]

2.2.1.4 SNAP-17

The SNAP-17 RTG was a power source designed to use a Strontium-90 Fluoride (SrF_2) heat source to provide power for the Medium Altitude Communications Satellite (MACS) System. MACSs were envisioned to circle the Earth in a Circular polar orbit of 5000 nautical miles and have all its power supplied by the SNAP-17 RTG. Two study contracts were issued for this system. The SNAP-17A generator concept was developed by the Martin Company Nuclear Division and used SiGe thermoelectric modules built by RCA. The SNAP-17A had a net power generation of 32 Watts and was expected to generate 29.5 Watts two years after launch and 26 Watts five years after launch. [19] The SNAP 17B generator was designed by General Electric with 3M producing the Lead Telluride thermoelectric generators. [20]

2.2.1.5 SNAP-29

The SNAP-29 power system design was to be integrated with a spacecraft. The concept is illustrated in Figure 2.13. This design comprised subsystems that included a heat source, a thermoelectric module, thermal control, heat rejection, and mounting structures. The primary objectives of the program were:

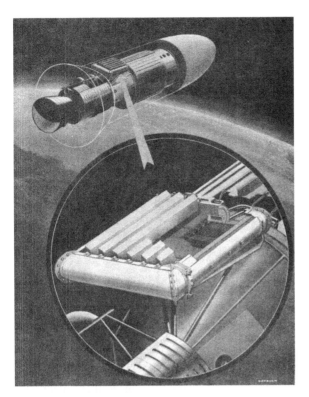

Figure 2.13 Illustration of the SNAP-29 concept showing integration with a spacecraft. [9] US Department of Energy.

- to operate for 144 days
- use a modular design with a fuel capsule that would remain intact during re-entry accidents and meets safety criteria.

The SNAP-29 was designed to produce a nominal electrical power output of 500 watts at EOM of 144 days after launch using polonium-210, at an overall weight of ~227 kg (500 lbs) or less. The SNAP-29 system was not developed into a flight unit. [9]

2.2.1.6 Selenide Isotope Generator

The Selenide Isotope Generator (SIG) was a power source proposed for the Galileo mission. It combined a modified MHW heat source similar to those used on the Voyager spacecraft coupled with selenide thermoelectric materials and a copper-water heat pipe radiator. The SIG system had a predicted thermoelectric efficiency of ~11% and system efficiency of 9%. Total system mass was 46.5 kg (102.5 lb) and would produce >214 W_e at beginning of life. The thermoelectric converters used Copper-Silver Selenide ($Cu_{1.97}Ag_{0.03}Se_{1+y}$) for the p-leg and Gadolinium Selenide ($GdSe_x$) for the n-legs. [21]

Teledyne Energy Systems and 3M were the two key contractors on the SIG program. Teledyne was responsible for systems integration, and 3M was responsible for the design, manufacture, and test of SIG thermoelectric converters. Reports from 3M and JPL in the 1970s noted that the selenide thermoelectric materials were selected over silicon germanium due to higher efficiency. However, the increase in selenium concentration caused problems during development. Mechanical property degradation and sublimation increased with higher concentrations of selenium, and caused changes in most of the properties of the thermoelectric material. These caused changes in most of the properties of the thermoelectric material due to the change in selenium composition. Challenges in developing consistent thermoelectric couples that would survive the mission life led to the program being cancelled in January 1979. [21]

2.2.1.7 Modular Isotopic Thermoelectric Generator

The Modular Isotopic Thermoelectric Generator (MITG) was designed around the GPHS heat source with each 'slice' of the MITG built around a single GPHS module. The program began in 1980 when DOE requested the Fairchild Space and Electronics Company to design a new RTG using new materials and improved design techniques. It used a multi-couple developed by the Syncal Corporation. [3,21,22]

Initial concepts of the MITG had 12 power modules containing 12 GPHS modules and would generate ~280 We. This compared favorably to the 18 GPHS modules in the GPHS RTG. Thermoelectric couples cracked during testing, and the thermoelectric material efficiency was less than initially expected. In September 1983, the MITG program ended and was rolled into the MOD-RTG program. [3,22]

2.2.1.8 Modular RTG

The goal of the MOD-RTG program was to develop a modular RTG ground demonstration system for testing. Main program directives were a specific power greater than 7.7 W/kg, a storage life of 3 years, and an operational life of 5 years. The program began in October 1983. The preliminary design review for the Ground Demonstration System occurred in March 1984 for major components and in July 1984 for the Ground Demonstration System and the Electric Heat Source. A final design review occurred in June 1985, which led to the fabrication of the Ground Demonstration System. The flight design used 18 GPHS modules, with a thermal output of between 4468 to 4673 Watts and an electrical converter efficiency between 7.0 and 7.7%. The MOD-RTG program encountered issues with mechanical and electrical shorting problems, fast thermoelectric degradation, and performance issues. The Ground Demonstration System was suspended in mid 1986 to focus on solving issues with multi-couples. The program ended in late 1992. [22]

2.3 Terrestrial RTGs

2.3.1 SNAP Terrestrial RTGS

2.3.1.1 SNAP-7

The SNAP-7 RTGs were deployed for maritime, Antarctic, ocean acoustic, and navigational applications, along with providing power to oil platform beacons. The SNAP-7D RTG is an example used for NOMAD Class maritime weather buoys, deep sea sonar, and light buoy navigation and is depicted in Figure 2.14. It used a strontium-90 titanate heat source and produced 10 to 30 watts of electricity. The SNAP-7 D used lead-telluride thermoelectric couples for power conversion and had a ten-year operating lifetime. An inert gas was used to fill the converter housing to prevent sublimation. That also served to minimize oxygen infiltration and prevent oxidation of the heat source and thermoelectric couples. The generator used depleted uranium for radiation shielding for biological protection and meeting shipping and handling requirements. [23]

Figure 2.14 Concept of a SNAP-7D thermoelectric generator powering a US Navy floating weather station. [16] Credit: [16] Cull (1989) US Department of Energy.

2.3.1.2 SNAP-15

The SNAP-15 series of RTGs were plutonium-238 fueled RTGs that produced a range of designs with electrical production less than 1 W at 4.5 Volts and five-year lifetimes. The thermoelectric couples used Tophel Special for the p-leg and Cupron Special for the n-leg. Figure 2.15 shows a cutaway of a SNAP-15 system. [24]

2.3.1.3 SNAP-21

The SNAP-21 generator was specifically developed for undersea applications and was the first radioisotopic generator for this application. [25] It used thermoelectric couples to produce 10 watts of electricity for undersea applications and an enhancement to this RTG was found to deliver 20 watts. It used strontium-90 as a heat source. Seven generators of the 10 watt design were fabricated and 5 were deployed. Figure 2.16 shows a cutaway of the SNAP-21 system. [25]

2.3.1.4 SNAP-23

The SNAP-23 RTG was developed for use on land and sea. [29] It used a strontium-90 heat source and generated 60 watts of DC power. It was a scaled-up version of the SNAP-21 generator that also attempted to develop a more efficient segmented couple and a five-year design life. [26, 29] The SNAP-23 was designed for remote weather stations. [29]

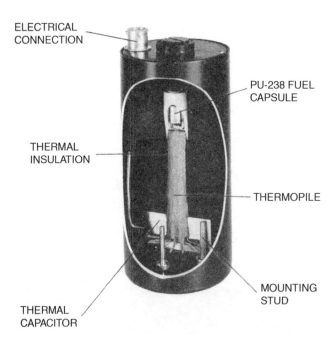

Figure 2.15 Model of SNAP-15 power source in cutaway. [18] Credit: [18] US Department of Energy..

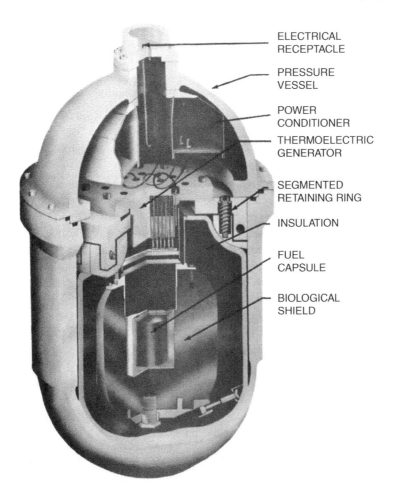

ELECTRICAL
RECEPTACLE

PRESSURE
VESSEL

POWER
CONDITIONER

THERMOELECTRIC
GENERATOR

SEGMENTED
RETAINING RING

INSULATION

FUEL
CAPSULE

BIOLOGICAL
SHIELD

Figure 2.16   Cutaway view of the SNAP-21 system. [19] US Department of Energy and Manufacturing Company.

Longer-term goals of the SNAP-23 program were to develop terrestrial RTGs with lifetimes of 10 years and high reliability. These power sources were envisioned to be superior to systems available at the time in terms of unit power cost, weight, and size. [23]

2.3.2   Sentinel 25 and 100 Systems

The Sentinel RTGs were powered by between 0.54 kg and 1.79 kg (1.2 lbs and 3.9 lbs) of strontium-90, which has a half-life of 28 years and is a byproduct of reprocessing of spent nuclear fuel. The strontium fuel form is a solid ceramic of strontium titanate in disks about the size of hockey pucks.

The Sentinel 25A, E, and F models generated between 9 and 20 watts of electric power output. The Sentinel 100F model generated at least 100 watts of electrical power at its beginning of life, but actual output depended on ambient air temperature. [27]

Sentinel RTGs have been deployed to several locations, including:

- One Sentinel-8 RTG powered a metrological data collection system on San Miguel Island, California, USA. It was installed in November 1970.
- One Sentinel-25A RTG, generating 25 watts, was installed on a small island in the Bering Sea. It generated power for oceanographic sensors.
- One Sentinel-25C1 RTG was placed on the San Juan Seamount at a depth of 2200 feet in June 1969. The experiment operated successfully for 5.6 years until it failed in January 1976. The system was recovered in April 1976 and tests showed the RTG was still operating after being underwater for nearly 7 years.
- Three Sentinel-25D RTGs were placed 11 miles offshore from Panama City, Florida, USA, to power surface wave gauges and a telemetering system to transfer data to shore. The units were placed in series to generate 60 watts of power at 3 volts.
- The Federal Aviation Administration's Lake Clark Communication Link Project installed five Sentinel-25F units in October 1977. The RTGs powered UHF relay stations to provide communications through the Lake Clark Pass, Alaska, USA. Each of the units produced 28 to 30 watts DC at 2.8 to 3.1 volts.
- One Sentinel-25F RTG was deployed for a Powered Polar Automated Weather Station (PAWS) on Minna Bluff, Antarctica from December 1976 to January 1977. It was removed due to storm damage.
- One Sentinel-25F RTG, generating 25 watts at 2.8 volts, was used to power a data buoy near Eleuthera Island, Bahamas, in February 1976.
- One Sentinel-100F was installed on Eleuthera Island, Bahamas, in December 1974. It generated 125 watts of power at 9 volts. It provided power for a quartz crystal clock that provided timing for a data acquisition system.

2.3.3 Sentry

The Sentry RTG was heated by strontium-90 and was deployed 700 miles from the North Pole on Axel-Heiberg Island, Canada, in August 1961. It powered an automatic weather station for four years. The RTG was returned to the United States in 1965 due to technical issues with the electronics in the weather station. The generator was still operating at the time of removal. [29]

2.3.4 URIPS-P1

The URIPS-P1 RTG produced 1 watt of electricity from a strontium-90 titanate heat source enclosed in a stainless steel clad. A Hastelloy C alloy enclosed each heat source to protect them from the marine environment. The Aerojet-General

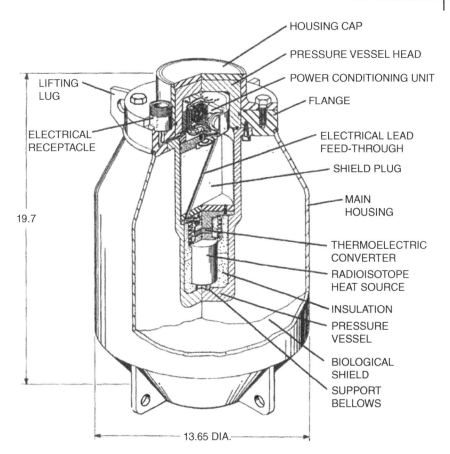

HOUSING CAP

PRESSURE VESSEL HEAD

POWER CONDITIONING UNIT

LIFTING LUG

FLANGE

ELECTRICAL RECEPTACLE

ELECTRICAL LEAD FEED-THROUGH

SHIELD PLUG

MAIN HOUSING

19.7

THERMOELECTRIC CONVERTER

RADIOISOTOPE HEAT SOURCE

INSULATION

PRESSURE VESSEL

BIOLOGICAL SHIELD

SUPPORT BELLOWS

13.65 DIA.

Figure 2.17 Cutaway view of URIPS power supply. [31] Credit: US Navy.

Corporation designed URIPS-P1 RTGs. The URIPS-P1 RTGs survived at a depth of 16,100 to 16,200 feet in the Pacific Ocean. [31] A cutaway view of the URIPS RTG is in Figure 2.17.

A URIPS-8 RTG was used to power an automatic weather station at Marble Point in Antarctica. It was installed in January 1976 to power transmitters of operational data and weather data to the NIMBUS F meteorological satellite. McMurdo Station in Antarctica took delivery of the RTG for storage in October 1977 because of issues with electronics in the weather station.

2.3.5 RG-1

General Atomics built the RG-1 RTG and used a strontium-90 titanate heat source enclosed in a stainless steel clad. The clad was encapsulated in a Hastelloy C alloy to provide protection from seawater corrosion. One RG-1 RTGs was tested at a

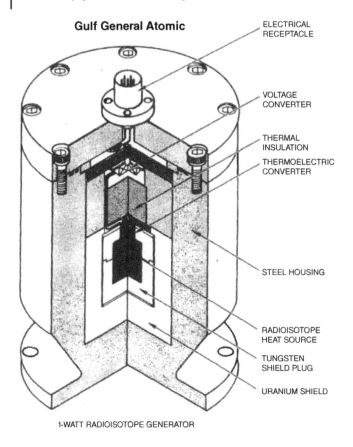

Gulf General Atomic

ELECTRICAL RECEPTACLE

VOLTAGE CONVERTER

THERMAL INSULATION

THERMOELECTRIC CONVERTER

STEEL HOUSING

RADIOISOTOPE HEAT SOURCE

TUNGSTEN SHIELD PLUG

URANIUM SHIELD

1-WATT RADIOISOTOPE GENERATOR

Figure 2.18 Cutaway view of RG-1 power system. [31] Credit: US Navy.

depth of 10,344 feet in the Atlantic Ocean and the second at a depth of 10,860 feet. [31] A cutaway view of the RG-1 is in Figure 2.18.

2.3.6 BUP-500

The BUP-500 RTG was designed by Teledyne Energy Systems as a demonstration using strontium-90 from the Hanford, Washington's Waste Encapsulation and Storage Facility. The goal of this project was to find beneficial uses for fission products and create a secure and constant power supply for national security applications. The design life was 5 years. The strontium-90 was in fluoride form and contained in a Haynes-25 liner. The BUP-500 generated 610 W_e at beginning of life. The lead/tellurium alloy TE 1006 and an alloy of tellurium, antimony, germanium, and silver alloy (TAGS-85) were the thermoelectric materials. [32]

2.3.7 Millibatt-1000

Two Millibatt-1000 RTGs were integrated with deep ocean transponders and dropped into the South Atlantic Ocean in February and March 1977. One of the RTGs powered a transponder for 10 years. The Millibatt-1000 deployed in February 1977 had a higher-than-expected descent rate and was damaged when it impacted the bottom of the ocean at a depth of ~3.4 km (2.1 miles). [28] The second system, deployed in March 1977, reduced the descent rate by attaching a structural plate under the RTG and was deployed successfully. The initial deployment was successful, but the transponder gave varying signals, which may have been the result of transponder issues or problems with the sling between the RTG and the transponder. [28,33] Plans for four additional systems of this type were cancelled due to these issues. [33]

2.4 Conclusion

Radioisotope systems have powered terrestrial and space missions since the 1960s. Their long endurance has helped to enable applications in remote locations and extreme temperatures. The development of new RTG concepts and designs has led to improvements in safety, specific power, system lifetime, and efficiency. Development of RTG systems with higher power conversion efficiencies is currently in work, which could further extend the robotic and human exploration of the solar system.

References

1 Lipp, J.E. and Salter, R.M. (1954). Project Feed Back Summary Report. *R-262*.

2 Voss, S. (1984). SNAP–Reactor Overview. *AD-A146 831. AFWL-TN-84-14*. Air Force Weapons Laboratory, Kirtland AFB.

3 Hula, G. (2015). Atomic Power in Space II: A History of Space Nuclear Power and Propulsion in the United States. *INL/EXT-15-34409*. Idaho National Laboratory.

4 Angelo, J.A. Jr. and Buden, D. (1985). *Space Nuclear Power*. Orbit Book Company: Malabar, FL.

5 Kardatzke, O.C. (1996). Dynamic Testing of SNAP-III Radioisotope Thermoelectric Generator. *NASA TM X-55603*. Goddard Space Flight Center. https://ntrs.nasa.gov/search?q=TM%20X-55603 (accessed 29 November 2022).

6 Bennett, G.L. (2006). Space Nuclear Power: Opening the Final Frontier. In: *4th International Energy Conversion Engineering Conference and Exhibit (IECEC)*, San Diego, California, (26–29 June 2006), 2006–4191. AIAA.

7 Fihelly, A.W., Berkow, H.N., and Baxter, C.F. (1968). Goddard Space Flight Center Publication No. *X-450-68-286.*

8 https://commons.wikimedia.org/wiki/File:Cutdrawing_of_an_SNAP_19_RTG. jpg (accessed 29 November 2022).

9 SNAP-29 power supply system. Ninth quarterly progress report. USA. https://doi. org/10.2172/4232480.

10 Schmidt, G.R., Sutliff, T.J., and Dudzinski, L.A. (2011). Radioisotopes – Applications in Physical Sciences, *Radioisotope Power: A Key Technology for Deep Space Exploration,* (ed. N. Singh). InTech. ISBN: 978-953-307-510-5.

11 Cull, T.A. (1989). General Purpose Heat Source Development: Extended Series Test Program Large Fragment Tests. *LA-11597-MC.*

12 Lee, Y. and Bairstow, B. (2015). *Radioisotope Power Systems Reference Book for Mission Designers and Planners.* JPL Publication.

13 Great Exploration Revisited: New Horizons at Pluto and Charon, (2021). http:// pluto.jhuapl.edu/News-Center/News-Article.php?page=20210714 (accessed 29 November 2022).

14 Casani, J.R. (2001). Report of the RPS Provisioning Strategy Team.

15 Mars 2020 Perseverance Launch Press Kit. National Aeronautics and Space Administration.

16 https://www.energy.gov/ne/articles/nuclear-power-system-delivered-florida-nasa-s-perseverance-rover (accessed 29 November 2022).

17 Morse, J.G. (1963). Energy for remote areas. *Science*, New Series 139 (3560): 1175–1180.

18 Snap 13: Thermionic Development Program Generator Fueling Report, (1967). USA. https://doi.org/10.2172/4655700

19 Streb, A.J. (1964). Snap 17A System, Phase I Final Summary Report. Part 1. USA. https://doi.org/10.2172/4634015

20 Snap-17B Power Supply Development Program (Phase I). Final Report, (1964). USA. https://doi.org/10.2172/4716713

21 Bennett, G. (2016). The selenide saga: a contribution toward a history of the selenide isotope generator. *14th International Energy Conversion Engineering, Propulsion and Energy Forum.* (25-27 July 2016).

22 Modular Radioisotope Thermoelectric Generator (RTG) Program. Final technical report, (1992). USA. https://doi.org/10.2172/10151769

23 Shor, R. et al. (1971). *Strontium-90 Heat Sources.* Oak Ridge National Laboratory, Isotopes Information Center.

24 Baker, F.S., Campana, R.J., and Sargent, W.S. (1967). Snap-15A Thermoelectric Generator. Final Report. USA https://doi.org/10.2172/4472063

25 Wickenberg, R.F., McArthur, W.A., and Pettman, R.E. (1968). Snap-21 Program, Phase II. Thermoelectric Generator Performance Evaluation and Design Description. USA. https://doi.org/10.2172/4802495

26 SNAP-23A Program Thermoelectric Converter Development Program Final Report, (1970). *Report No. MMM-4187-0001.*

27 US Congress, Office of Technology Assessment, Power Sources for Remote Arctic Applications, (1994). *OTA-BP-ETI 129.* Washington, DC

28 Naval Nuclear Power Unit Port Hueneme (1978). *Radioisotope Thermoelectric Generators of the US Navy*, vol. 10. California: Final report 1 Jan 1976–1 Jul 1978.

29 United States Atomic Energy Commission (1966). *Major Activities in the Atomic Energy Programs, January–December 1965.*

30 Corliss, W.R. and Mead, R.L. (1971). Power From Radioisotopes. In: *United States Atomic Energy Commission Division of Technical Information*, 74–169081. Available at: https://www.osti.gov/includes/opennet/includes/Understanding%20the%20Atom/ Power%20from%20Radioisotopes%20V.3.pdf (Accessed 27 July 2020).

31 Goodfellow, S.T. (2018). Request for technical assistance in the abandonment of radioisotope thermoelectric generators in situ at the bottom of the ocean. Ser *N452-RCO/18U132434.* Department of the Navy Letter.

32 Vogt, J.F. and McMullen, W.H. (1985). *The BUP-500 Radioisotope Thermoelectric Generator*, 859461. Society of Automotive Engineers.

33 Rosell, F.E. Jr. and Vogt, J.F. (1978). Recent terrestrial and undersea applications of radioisotope thermoelectric generators. Proceedings of the 28th Power Sources Symposium (12–15 June 1978).

3

US Space Flights Enabled by RTGs

Young H. Lee and Brian K. Bairstow

Jet Propulsion Laboratory/California Institute of Technology, Pasadena, California

For more than 60 years, the US has produced a string of successful space flights using RTGs with no mission losses caused by an RTG failure. Those flights, their missions, and the means of how RTGs enabled the missions are presented.

3.1 SNAP-3B Missions (1961)

3.1.1 Transit 4A and Transit 4B

The US Navy sponsored the Transit satellite program developed by the US Defense Advanced Research Projects Agency (DARPA) and the Johns Hopkins Applied Physics Laboratory (APL). The satellites's signals were used for navigation and hydrographic and geodetic surveying. These missions furthered our understanding of ionospheric refraction of radio waves and of the Earth's shape and gravitational field. The Transit spacecraft were precursors to the Global Positioning System (GPS) [1,2], and the Transit 4A and 4B missions marked the first successful use of RTGs in space.

Solar power and batteries had powered previous Transit spacecraft. However, the use of RPS technology had been under consideration from the start of the program because of concerns that the hermetic seals on the rechargeable spacecraft batteries would not last for the five-year mission requirement. The Transit program worked with the Atomic Energy Commission (AEC) to introduce the SNAP-3 RTG as an auxiliary power source. The first SNAP-3 RTG used a polonium heat source. A trade study over the use of strontium-90 or plutonium-238 over polonium-210 found strontium-90 was not a practical option because of the additional mass required to shield the isotope's radiation emissions. The design of the RTG was changed from using ^{210}Po to ^{238}Pu and became the SNAP-3B. [2]

The Technology of Discovery: Radioisotope Thermoelectric Generators and Thermoelectric Technologies for Space Exploration, First Edition. Edited by David Friedrich Woerner.

Figure 3.1 Transit 4A spacecraft with companion payloads before launch. Transit 4A is the cylindrical spacecraft at the bottom of the stack. [1], NASA.

The Transit 4A satellite (Figure 3.1) launched on 29 June 1961 and Transit 4B launched on 15 November 1961. Each spacecraft carried one SNAP-3B designed to provide 2.7 W_e of power at the beginning of mission (BOM) to power the crucial crystal oscillator used for Doppler shift tracking. [2]

The RTG on Transit 4A had an initial peak power of 2.8 W_e (Figure 3.2). Unfortunately, there are only 17 days of voltage telemetry data for Transit 4A because of the failure of the spacecraft's telemetry transmitter. However, the other transmitters on the spacecraft enabled confirmation that the RTG was still operating throughout the mission's lifetime. The mission continued for over 15 years, despite the failure of the oscillator circuit's voltage regulator and degradation of the spacecraft's solar cells, until the mission ended in 1976. [2,4]

The RTG on Transit 4B had an initial peak power of 3.1 W_e, and operated from launch on 15 November 1961, until 6 June 1962, when two power excursions occurred, most likely because of a capacitor failure in the power-conditioning subsystem downstream of the RTG. The solar cells suffered severe degradation after a high-altitude nuclear test conducted on 9 July 1962, and the spacecraft ceased transmitting on 2 August 1962. Contact with the spacecraft resumed in 1967 and was lost again in April 1971. However, based on the telemetry data, the RTG was still functioning. [4]

3.2 SNAP-9A Missions (1963–1964)

3.2.1 Transit 5BN-1, 5BN-2, and 5BN-3

The Transit 5BN-1, 5BN-2, and 5BN-3 missions followed the Transit 4A and 4B missions in the Transit program, continuing with the same objectives. Because of the success of the SNAP-3B RTGs, these missions were the first to rely solely on an

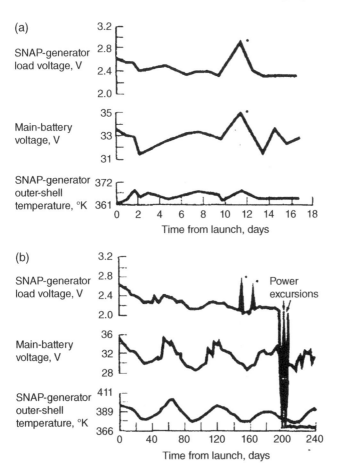

Figure 3.2 Transit 4A (a) and Transit 4B (b) SNAP-3B telemetry data. Voltage (V) changes (marked by asterisks) were due to load data. Power excursions probably resulted from a capacitor failure (not on the RTG). °K is degrees Kelvin. [3]

RTG as the primary power source. The SNAP-3B RTG's greater inherent immunity to solar radiation was a critical factor in the decision to develop the SNAP-9A as the solar cells on the previous missions had suffered due to radiation from high-altitude nuclear testing. [2,4]

Transit 5BN-1 launched on 28 September 1963, and Transit 5BN-2 launched on 5 December 1963. Transit 5BN-3 launched on 21 April 1964, but the launch was aborted at an altitude of 1600 km over the South Pole, where it reentered the atmosphere. The [238]Pu heat source was designed to burn up so that the radioisotope would disperse in the atmosphere, to distribute the potential dose below the level that would cause a health hazard. Indeed, subsequent atmospheric testing using balloons confirmed that the fuel dispersed as expected. [2]

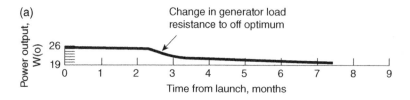

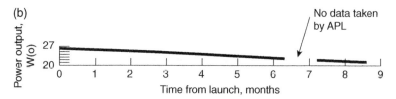

Figure 3.3 Transit 5BN-1 (a) and Transit 5BN-2 (b) power (W(o)) histories (smoothed data). [3]

One SNAP-9A RTG, which provided continuous power for five years in space after one year of storage on Earth, powered each spacecraft. As shown in Figure 3.3, the BOM power was 26.8 W_e, nearly ten times the power of the SNAP-3B RTGs, and sufficient to power the spacecraft without solar panels. [2]

Transit 5BN-1 operated well for two months, at which point it suffered a short-circuit either in a wiring harness or in the electronics, resulting in an excess load that caused off-nominal performance. Final telemetry data were received on 1 June 1964. After that, Transit 5BN-2 operated for nearly a year until solar heating caused faults in the satellite time-keeping system that made it inadequate for navigation. However, RTG telemetry data continued to be transmitted by the spacecraft until June 1970. [4]

The missions succeeded in demonstrating the satisfactory operation and long-life potential of the SNAP-9A RTG even though the spacecraft carrying them did not operate for their five-year design lifetimes. [4]

3.3 SNAP-19 Missions (1968–1975)

3.3.1 Nimbus-B and Nimbus III

The Nimbus series of satellites launched between 1964 and 1978 for meteorological research and as technology refinement platforms. At the end of the Nimbus program, the refined technology was transferred from NASA to the US National Oceanic and Atmospheric Administration (NOAA).

Nimbus-B was an Earth-orbiting weather satellite designed to take day and night temperature measurements at various altitudes and measure solar radiation

Figure 3.4 Intact SNAP-19 fuel capsule from Nimbus B aborted launch, shown in foreground among debris on Pacific Ocean floor. US Department of Energy, Wikimedia Commons, Public domain.

above the atmosphere. The Nimbus-B mission was NASA's first mission to carry an RTG. The Nimbus-B satellite launched on 18 May 1968, using two SNAP-19 RTGs. Its launch lasted approximately 120 seconds before it was aborted. The mission ended with the spacecraft and RTGs in the Pacific Ocean, in the Santa Barbara Channel, off of the California coast (Figure 3.4). The RTGs operated as designed. They were not the cause of the aborted launch; the fuel containers were intact and recovered from the sea floor [5,6] and refurbished for the two SNAP-19B3 RTGs flown on the Nimbus III mission.

The subsequent Nimbus III mission (Figure 3.5), also denoted Nimbus-B2, was the first NASA spacecraft to orbit radioisotope power in space. Nimbus III provided the first orbital measurements of air pressure, day and night temperatures, solar ultraviolet radiation, the ozone layer, and sea ice. [1] The mission had three objectives regarding the SNAP-19B3 RTGs: to integrate an RTG with a complex NASA spacecraft, to assess the performance of an RTG in space over an extended period, and to use the power from the RTGs to supplement the Nimbus III power. [7]

The Nimbus III satellite was identical to Nimbus-B and launched on 14 April 1969, and operated until 22 January 1972. The spacecraft mass totaled 571 kg and included 2 SNAP-19B3 RTGs and solar panels. [6] The output of the RTGs amounted to 20% of the total spacecraft power, the remainder coming from the

Figure 3.5 Nimbus III spacecraft with 2 SNAP-19 RTGs visible on the left. [1], NASA.

solar arrays. Without the RTGs, the mission would have been short the power needed to continuously operate its atmospheric-sounder experiments just two weeks after launch. [2]

As shown in Figure 3.6, the RTGs produced 56.4 W_e at launch and 47 W_e after one year. [8] The degradation rate was higher than expected based on laboratory testing of electrically heated generators. This increased degradation rate was

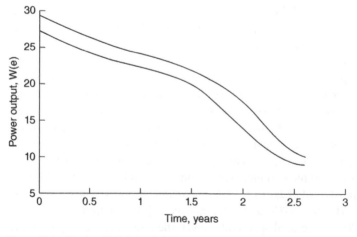

Figure 3.6 Nimbus III SNAP-19 RTGs power output (W(e)) (smoothed data). [3]

likely due to gasses leaking from or forming within the generator: the leakage of the argon cover gas from the generator into space, the introduction of helium produced by the decay of the fuel, and the carbon dioxide and carbon monoxide produced from oxygen released from the fuel. After 2.5 years, the units experienced a sharp degradation in performance attributed to the sublimation of thermoelectric material and degeneration of the hot junction bond because of depletion of the internal cover gas used to slow sublimation. [2] The possible sources of degradation caused by leakage of beneficial gases or production of detrimental gases were designed out in subsequent SNAP-19 designs. [3]

The successful demonstration of the RTG on Nimbus III encouraged NASA to use the SNAP-19 on the Pioneer and Viking missions. [2]

3.3.2 Pioneer 10 and 11

The Pioneer 10 and 11 spacecraft were twins that performed different missions. Pioneer 10 was the first spacecraft to fly beyond Mars' orbit, through the asteroid belt, and visit Jupiter (Figure 3.7). The mission's science goals were to explore the interplanetary medium beyond Mars, to investigate the asteroid belt both from a scientific standpoint and as a potential hazard to future missions, and to explore the environment of Jupiter. The spacecraft took the first close-up images of Jupiter, charted the planet's radiation belts, located the planet's magnetic field, and performed a gravity assist maneuver. Pioneer 11 also performed a Jupiter

Figure 3.7 Pioneer 10/11, with 2 RTGs visible on each of 2 booms. [1], NASA.

flyby and a gravity assist maneuver and then became the first spacecraft to perform a Saturn flyby. It also returned the first close-up pictures of Saturn, discovered two small moons and a Saturnian ring, and charted the planet's magnetosphere and magnetic field. [1]

Pioneer 10 launched on 2 March 1972, while Pioneer 11 followed on 5 April 1973. Pioneer 10 performed its Jupiter flyby in December 1973, while Pioneer 11 performed its Jupiter flyby in December 1974, and its Saturn flyby in September 1979. [8] At launch, each spacecraft had a mass of 259 kg. [9]

Pioneer 10 and 11 were each powered by four SNAP-19 RTGs, with additional heat provided by 12 radioisotope heater units (RHUs). The RTGs were enhanced from the Nimbus III SNAP-19B3 RTGs, in both specific power and degradation rate. Each mission had to have at least 120 W_e at their Jupiter flyby. The spacecraft required 100 W_e to power all systems, including 26 W_e for the science instruments. [9] RTG power enabled these deep-space missions that were the first to explore beyond the orbit of Mars and reach the outer planets, as the solar array technology available at the time was not capable of delivering enough power at a solar distance of over five astronomical units (AU).

The Pioneer SNAP-19 RTGs produced an average of 40.3 W_e per RTG at BOM (Figure 3.8). At Jupiter flyby, the Pioneer 10 RTGs produced 144.0 W_e, while the Pioneer 11 RTGs produced 142.6 W_e, thus meeting the mission power requirements. For the extended mission to Saturn, the mission requirement was for

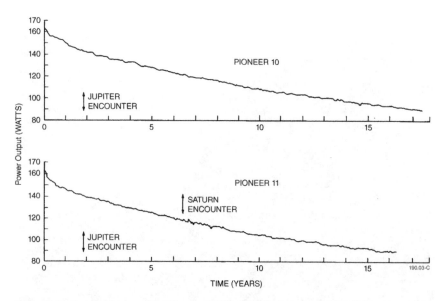

Figure 3.8 Power history of the Pioneer SNAP-19 RTGs (summed data). [11]

the RTGs to provide at least 90 W_e at Saturn, which the Pioneer 11 RTGs exceeded by producing 119.3 W_e. [8]

The RTGs continued to power their spacecraft for decades, even though the original lifetime requirement on the Pioneer SNAP-19 RTGs was 21 months of operation. Pioneer 10 continued its extended mission until March 1997, out to a distance of 67 AU from the Sun, and the last signal from the spacecraft was received on 22 January 2003. After that, the electrical power output of the RTGs had dropped too low to power the transmitter. The Pioneer 11 mission ended in October 1990 because of a failure in its communication system, yet the spacecraft was 40 AU from the Sun when it sent its last signal in November 1995. [8,9]

3.3.3 Viking 1 and 2 Landers

Viking 1 and Viking 2 were missions to Mars. Each comprised an orbiter and a lander (Figure 3.9). The mission goals were to take high-resolution images of the Martian surface, characterize the structure and composition of the atmosphere and surface, and search for evidence of life. Viking 1 was the first mission to land on Mars and return science data, and Viking 2 was also successful. The missions returned the first in situ images of the Martian surface. In addition, they found that carbon, nitrogen, hydrogen, oxygen, and phosphorus–all the elements essential to life on Earth–were present on Mars. [1]

Figure 3.9 Viking lander model—the two RTGs are covered by windscreens and are inside the lander's body. [10], NASA.

Viking 1 launched on 20 August 1975, and landed on the western slope of Chryse Planitia on 20 July 1976. Viking 2 launched on 9 September 1975, and landed at Utopia Planitia on 3 September 1976. [2] Each lander had a mass of ~600 kg.

Two SNAP-19 RTGs powered each Viking lander. The minimum RTG power requirement was 35 W_e for 90 days after an 11- to 12-month cruise phase. [3] The landers also carried four nickel-cadmium batteries used during daily peaks in activities and later recharged by the RTGs. In addition, the lander's design used the RTG's waste heat to maintain the temperature of the electronics to preclude any degradative effects from thermal cycling during each Martian diurnal cycle. [8]

The average power produced by the Viking SNAP-19 RTGs was 42.7 W_e at launch (Figure 3.10). The RTGs easily met the power requirement of 70 W_e for each lander at 90 days after landing, with Viking 1 producing 79 W_e and Viking 2 producing 82 W_e. Each lander operated for years beyond the 90-day original mission, which would not have been possible without the electrical and thermal power from the RTGs. Failures in other lander subsystems and human error caused the eventual end of the missions. NASA lost communications with the Viking 1 lander on 13 November 1982, while the Viking 2 lander was inadvertently shut down on 1 February 1980. The final power estimate for Viking 1 RTGs estimated the lander would have sufficient power for operation until 1994, which would have been 18 years longer than the original 90-day design point. [8]

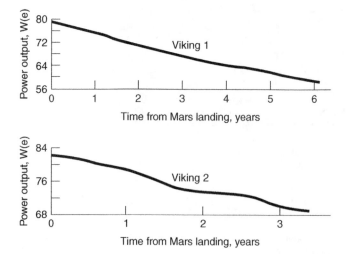

Figure 3.10 Power (W(e)) history of the Viking SNAP-19 RTGs (summed and smoothed data). [11]

3.4 SNAP-27 Missions (1969–1972)

3.4.1 Apollo 12–17

NASA requested the AEC develop a new RTG, the SNAP-27, to power the Apollo Lunar Surface Experiments Packages (ALSEPs) because the required power levels would have required multiple SNAP-19 RTGs, which would have required more deployments and radioactive exposure for the Apollo crews. [2] For each successful Apollo landing, starting with the Apollo 12 mission, US astronauts left an ALSEP on the surface of the Moon to perform long-term science experiments and transmit their data back to Earth. The ALSEP's instruments included passive and active seismometers, magnetometers, solar wind spectrometers, supra thermal ion detectors, and heat flow experiments. The seismometers collected the Moon's interior structure data by observing meteoroid impacts and moon quakes. The magnetometers discovered a faint magnetic field, surprising as the Moon has no active internal dynamo. The results of the heat flow experiments set limits on the thermal history of the Moon and on lunar radioactivity, as the decay of thorium, uranium, and potassium are the long-term sources of internal heating of the Moon. Other instruments tracked changes in solar and terrestrial magnetic fields and the composition of the lunar atmosphere. [1,12]

Apollo 12 launched in November 1969, while NASA launched the last Apollo mission in December 1972. In total, NASA deployed five ALSEPs on the Moon (Figure 3.11).

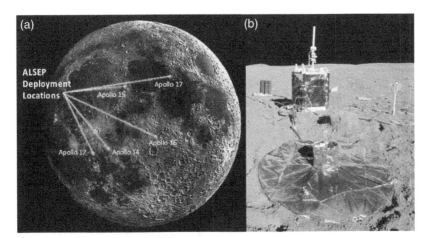

Figure 3.11 (a) Apollo landing sites where ALSEPs were deployed. Credit: NASA (b) Apollo 16 ALSEP, with RTG visible as the finned, grey, rectangular device in the upper left. The passive seismic experiment is in the foreground, surrounded by its circular reflective thermal skirt, while the central station, in square, reflective blankets, is between the RTG and the seismometer. [12], NASA.

A single SNAP-27 RTG powered each ALSEP. Had NASA chosen to power the ALSEPs with solar arrays and batteries, the Agency would have taken the risk of accumulated lunar dust on the arrays stopping the experiments. The SNAP-27s enabled multi-year science experiments that had to survive many lunar nights that were 14-Earth-days long. The power requirement for Apollo 12 through 16 was for the RTGs to each provide 63.5 W_e at 16 V DC after one year. An example from Apollo 17: the RTG delivered 69 W_e after two years. [3] All five of the ALSEPs and their RTGs operated for five or more years, well past their design lifetime. There is evidence in the historical record that the RTGs may have been degrading faster on the Moon than expected because of higher RTG temperatures than predicted, temperatures above the qualification temperatures for the SNAP-27 RTGs.

The SNAP-27 RTGs (Figure 3.12) met each Apollo mission's power and lifetime requirements, with degradation matching performance predictions. [2] All five ALSEPs operated until all stations were shut down on 30 September 1977, for budgetary reasons. [3]

The Apollo 13 mission also carried a SNAP-27, which never made it to the Moon's surface due to an in-flight accident that required the astronauts to return to the Earth without landing. The SNAP-27 fuel cask was on the attached lunar module during re-entry into the Earth's atmosphere. NASA selected the re-entry trajectory and orientation to ensure proper disposal of

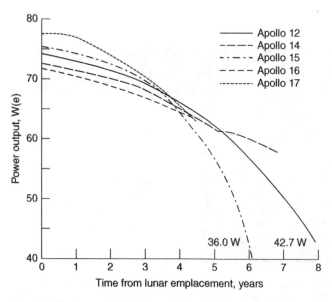

Figure 3.12 Power (W(e)) history of the ALSEP SNAP-27 RTGs (smoothed data). [3]

the cask, as per the pre-launch planning for this type of abort event. The lunar module detached and broke up on re-entry, as expected, while the graphite-encased fuel cask remained intact and ended up 6-9 km deep in the Tonga Trench as intended. Subsequent testing detected no increase in background radiation in the area. [2]

3.5 Transit-RTG Mission (1972)

3.5.1 TRIAD

The Triad spacecraft, depicted in Figure 3.13, was a navigational satellite designed to test improvements to the earlier Transit satellites. It was part of the Transit Improvement Program, which had the objective of developing a radiation-hardened satellite that could maintain its position for five days without communicating with the ground. [2]

The mission launched on 2 September 1972, powered by a single Transit-RTG, with auxiliary power from solar panels and a 6 Amp-hr battery. The RTG generated 35.6 W_e at BOM, dropping to 35.4 W_e after 30 days. [2,4]

The mission met the short-term objectives, including the initial demonstration of the RTG's performance. Unfortunately, the RTG's power data monitoring ended on 2 October 1972, due to a telemetry-converter failure. The RTG continued operating, and the satellite remained operational and returned magnetometer data for at least a decade.

Figure 3.13 Triad spacecraft. Credit: US Navy.

3.6 MHW-RTG Missions (1976–1977)

3.6.1 Lincoln Experimental Satellites 8 and 9

Lincoln Experimental Satellites 8 and 9 (LES 8 and 9) were notable RTG-powered US spacecraft (Figure 3.14). The Massachusetts Institute of Technology (MIT) Lincoln Laboratory built the LES 8 and 9 for the US Air Force to demonstrate advanced strategic communications links. On 15 March 1976, they launched together to geosynchronous orbits. [2] Each spacecraft had a launch mass of 450 kg.

Two Multi-hundred Watt (MHW) RTGs powered each spacecraft, with power requirements for each RTG to provide 125 W_e at 26 V at five years after launch. The MHW-RTGs were developed primarily for the upcoming Voyager missions, yet the LES 8 and 9 spacecraft launched with MHW-RTGs before the Voyagers. The designers of the Earth-orbiting LES 8 and 9 spacecraft selected the MHW-RTGs because they offered the advantages of continuous power, Figure 3.15, during regular 70-minute eclipses along with improved physical robustness compared to solar arrays. [2,3]

All four MHW-RTGs met their power and lifetime requirements. LES-8 operated until 2 June 2004, when the command system began to malfunction. The spacecraft was shut down. LES-9 operated for over 44 years, until the gradual power degradation, along with the loss of the primary S-band telemetry data in early 2020, led to the decommissioning of the satellite on 20 May 2020. [14]

Figure 3.14 LES 8 and 9 deployed in geosynchronous orbit. [2] Schmidt (2009), NASA.

(a)

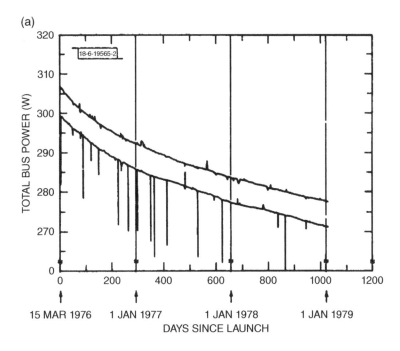

(b)

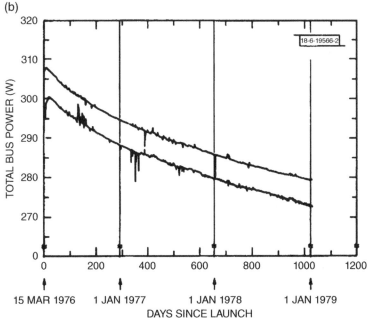

Figure 3.15 Daily maximum and minimum RTG output powers (W) for LES-8 during its initial 1,000 days in orbit, (a). [13] Daily maximum and minimum RTG output powers (W) for LES-9 during its initial 1,000 days in orbit, (b). [13]

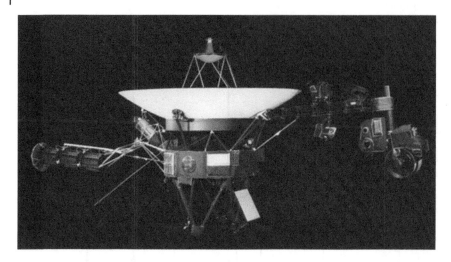

Figure 3.16 A Voyager spacecraft, with three MHW-RTGs end-to-end on the boom on the left. [15], NASA.

3.6.2 Voyager 1 and 2

The Voyager missions explored the outer solar system, with Jupiter, Saturn, Uranus, and Neptune flybys. These two spacecraft (Figure 3.16) made many significant discoveries, including Jupiter's rings, volcanos on Io, and geysers on Triton. [1] The Voyager missions explored 48 moons on top of visiting all four of the gas giant planets. [2]

Voyager 2 launched first on 20 August 1977, and Voyager 1 launched on 5 September 1977. The launched mass of each spacecraft was 815 kg, with 105 kg of that being the payload, which included cameras, spectrometers, photometric instruments, radio receivers, and fields and charged particles sensors. The power need of the payload was 90 W_e for science experiments and 10 W_e for heating of the spacecraft's electronics. [16]

On 25 August 2012, Voyager 1 became the first spacecraft to pass through the heliopause and enter interstellar space at 122 AU from the Sun. Voyager 2 crossed the heliopause on 5 November 2018. Both spacecraft were operational in 2021.

Three MHW-RTGs powered each spacecraft, producing 470 W_e at BOM (Figure 3.17). The mission requirement was for each RTG to produce 128 W_e at the end of the primary mission (EOM) 4 years after launch. [3] Forty-four years later, in 2021, the current power production was about 249 W_e for each spacecraft. [15] As the power decays, the mission team must choose which power loads to turn off, such as the heater for the Voyager 1 ultraviolet spectrometer instrument in

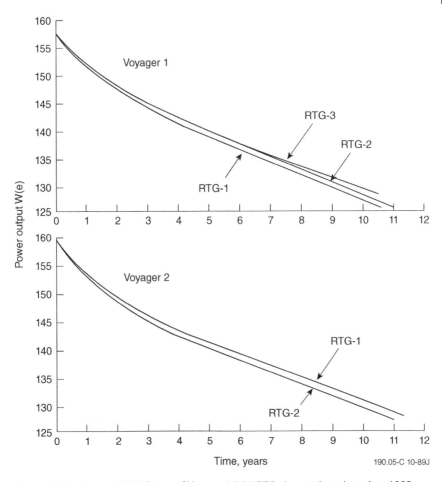

Figure 3.17 Power (W(e)) history of Voyager MHW-RTGs, incomplete data circa 1998. Note that telemetry was not operating for the 3rd RTG on Voyager 2. [11]

2012 and the heater for the Voyager 2 cosmic ray instrument in 2019. Voyager 2 continues to return data from five of its scientific instruments. [17]

RTG power enabled these missions, as solar array technology was still not at the point of making Jupiter missions feasible and missions to Uranus and Neptune were impossible without an RTG. Past 100 AU in the interstellar medium, the spacecraft now see less than 0.01% of the sunlight we receive on Earth. NASA expects both spacecraft to operate until ~2032, when there is likely to be inadequate power to operate the science instruments. [15]

3.7 GPHS-RTG Missions (1989–2006)

3.7.1 Galileo

The Jupiter Orbiter Probe was the first spacecraft sent to orbit Jupiter and study the planet, its moons, and its atmosphere and magnetosphere. NASA renamed the mission Galileo before liftoff. The mission made discoveries about Io's volcanism and plasma interactions with Jupiter, and found evidence for liquid water oceans under the ice on the moon Europa, and potentially the moons Ganymede and Callisto. In addition, the mission carried a battery-powered probe that took the first samples of a gas planet's atmosphere, measuring temperature, pressure, chemical composition, cloud characteristics, sunlight, heat, and lightning. [1]

NASA delayed the launch of the Galileo spacecraft from May 1986 because of the space shuttle Challenger accident in 1986. [3] Galileo launched from the Space Shuttle *Atlantis*, on 18 October 1989, and performed a Venus gravity assist and two Earth gravity assist maneuvers on its way to Jupiter, where it arrived on 7 December 1995. The dry mass of the orbiter (Figure 3.18) was 1,880 kg (wet mass was 2,223 kg), including 118 kilograms for the instruments. [18]

Figure 3.19 shows two power curves, one for each of the two General-Purpose Heat Source (GPHS) RTGs that powered Galileo producing 570 W_e at BOM and 490 W_e at arrival at Jupiter six years later. [18] The long-lived power generation allowed for three mission extensions from Galileo's primary mission of two years

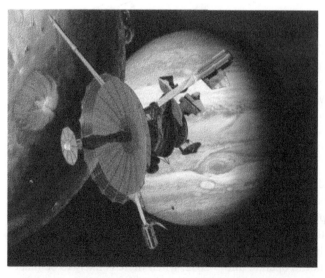

Figure 3.18 Galileo at Io, with Jupiter in the background. The two GPHS-RTGs are visible on separate booms. [18], NASA.

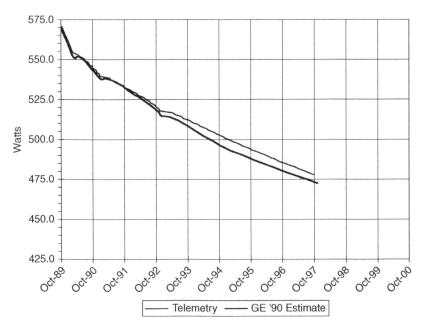

Figure 3.19 Galileo Total RTG Power (telemetry and predictions, GE '90 Estimate). [19]

in orbit to eight years in orbit. The spacecraft was running low on fuel after almost eight years after arrival and 35 orbits of Jupiter. Therefore, on 21 September 2003, to avoid any risk of contaminating any local moons with biological matter from Earth, the spacecraft was flown into the atmosphere of Jupiter to burn up. [2,3]

RTG power enabled the mission, as solar panels were not practical at Jupiter due to low solar insolation, cold temperatures, and high radiation. Galileo also carried 120 Radioisotope Heater Units (RHUs) with 1 W of heat per unit, which offset the need to use electricity to heat spacecraft components. 103 RHUs flew on the Galileo spacecraft, and 17 were aboard the atmospheric probe carried to Jupiter.

3.7.2 Ulysses

The International Solar Polar Mission (ISPM) was a joint effort between NASA and the European Space Agency (ESA), and was renamed the Ulysses mission. The mission used a gravity assist maneuver at Jupiter to bring the spacecraft out of the solar system's ecliptic plane and into a polar orbit around the Sun. The mission studied the heliosphere and made many important discoveries about the solar wind and the Sun's magnetic field. In addition, the mission returned findings on the three-dimensional character of galactic cosmic radiation and

Figure 3.20 Ulysses spacecraft. [20], NASA.

energetic particles from solar storms and the solar wind and made the first direct measurements of interstellar dust particles and interstellar helium atoms. [1]

Ulysses (Figure 3.20) launched on 6 October 1990, and performed its swing-by maneuver at Jupiter on 8 February 1992, which put the spacecraft into a polar orbit around the Sun with a period of 6.2 years. [20] The Space Shuttle Challenger accident delayed the Ulysses launch as it did the Galileo launch, from the original launch date of May 1986. As with Galileo, an RTG enabled the mission, as solar power was not workable at 5 AU from the Sun.

Early on, the joint NASA-ESA ISPM team requested a new, larger, more powerful RTG than the MHW-RTG for their spacecraft. This desire for more power may have influenced the GPHS-RTG design, which was already under development, to become the largest RTG ever built. GPHS-RTG used the same basic Si-Ge unicouple design as the MHW-RTG with a new housing, and most importantly, a new heat source with improved safety characteristics. A single GPHS-RTG powered Ulysses and produced ~282 W_e at liftoff (Figure 3.21). [2] The successful performance and long lifetime of the RTG enabled ESA to extend the mission four times. This extended mission duration allowed the spacecraft to complete close to three polar orbits. The decreasing power of the RTG and failure of the X-band radio degraded the spacecraft's ability to power science instruments and keep the

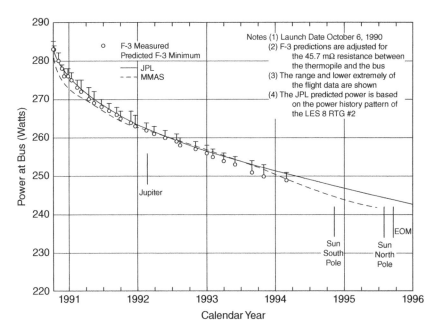

Figure 3.21 Measured and Predicted Power from the Ulysses GPHS-RTG. F-3 is the serial number of the RTG. mΩ stands for milli-Ohm. [21]

hydrazine fuel from freezing in the end. [2,3] As a result, the mission team ended the mission on 30 June 2009, when communications stopped. By that date, the hydrazine fuel onboard had frozen as the spacecraft was consuming more energy than the RTG could provide. [22]

3.7.3 Cassini

Cassini was the first spacecraft to orbit Saturn, where it studied the planet's moons, atmosphere, rings, and magnetosphere. Mission discoveries include Earth-like processes on Titan and plumes of icy material on Enceladus. The mission also had a battery-powered probe, Huygen, that was provided by the European Space Agency (ESA), and Huygen's landing on Titan marked the first landing in the outer solar system. [1]

Cassini launched on 15 October 1997, and after a 6.7-year transit that included Venus and Earth gravity assist maneuvers, Cassini entered orbit at Saturn on 1 July 2004. The spacecraft (Figure 3.22) had a dry mass of 2,100 kg. [23] As solar power was not a viable option at Saturn, the mission was enabled by the use of three high-power RTGs and 117 RHUs.

Figure 3.22 Cassini-Huygens at Saturn. Two of the GPHS-RTGs are visible on the left side of the closest end of the spacecraft opposite the antenna, while the third is visible in part on the right. [1], NASA.

Three GPHS-RTGs powered Cassini, making it the mission to fly the most plutonium and generate the most RTG power. The measured total power generated at launch was 887 W_e, which exceeded the power requirement of 826 W_e (Figure 3.23). This BOM power was higher per RTG than those RTGs dedicated to the delayed Galileo and Ulysses spacecraft because those missions flew with plutonium that had already experienced additional years of natural decay. [3] The power generation onboard the Cassini spacecraft after 20 years of operation was 633 W_e. [23]

After the primary mission of four years at Saturn, NASA approved extending the mission on 31 May 2008, for two years and later for another seven years. Its fuel dwindling, the mission had to be terminated for planetary protection reasons to prevent any chance of biological contamination of Saturn's moons, even though the spacecraft still had sufficient power to function and conduct science experiments. The spacecraft carried out its "Grand Finale" and performed its final science experiments by passing through Saturn's rings and entering its atmosphere on 15 September 2017.

3.7.4 New Horizons

The New Horizons mission carried out the first close look at Pluto and its moons. Besides taking the first clear images of the dwarf planet, the spacecraft characterized its global geology and morphology, mapped its chemical compositions, and

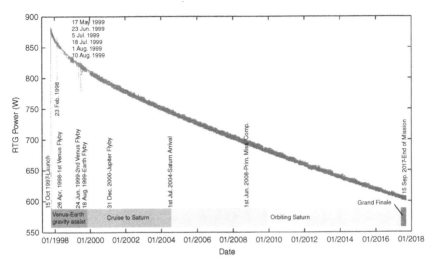

Figure 3.23 Cassini recorded RTG power output telemetry data over the entire mission between launch and EOM. [24]

characterized its atmosphere. Three and one-half years after visiting Pluto, the spacecraft conducted the first investigation of a Kuiper Belt Object, 486958 Arrokoth, which characterized the geology and morphology and mapped the surface composition.

New Horizons (Figure 3.24) launched on 19 January 2006 as the first competitively selected—rather than directed—NASA mission powered by an RTG. In 2007, it performed a gravity assist maneuver at Jupiter, while taking science measurements. The closest approach to Pluto was on 14 July 2015, after 9.5 years of transit. The extended mission flyby of Arrokoth took place on 1 January 2019. New Horizons continues to observe many Kuiper Belt Objects (KBOs) from a distance, and there is potential for a third encounter if the mission team can identify another KBO reachable by the spacecraft with its remaining fuel. The dry mass is 401 kg, including 30 kg of payload.

A single GPHS-RTG powers the spacecraft, a spare from the Cassini mission's provision. The initial fueling plan for the RTG included plutonium purchased from Russia that would be newly processed. However, due to fuel processing delays, it was fueled with only 9.75 kg PuO_2 (compared to 10.9 kg for Cassini), a mixture of new fuel and older fuel from Cassini's spare RTG. This fueling arrangement resulted in a ~15% lower power output at BOM, 245.7 W_e. [8,25,26] The RTG's power output has been decreasing by about 3.5 W_e per year and was about 190 W_e in January 2019 at the Arrokoth encounter. [27] More recently, the power decrease has been at 3.2 W_e per year, with the spacecraft reporting power output in January 2021 at 183 W_e (Figure 3.25). The spacecraft operates without a battery

Figure 3.24 Artist's rendition of New Horizons at Pluto, with GPHS-RTG in the left foreground. [1], NASA.

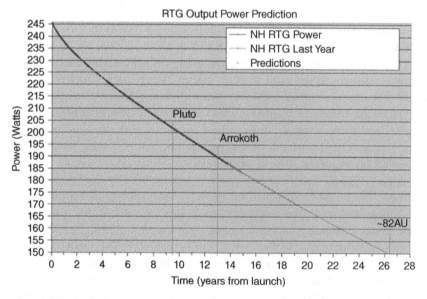

Figure 3.25 New Horizons (NH) Power History. Credit: NASA/Johns Hopkins University/ Applied Physics Laboratory.

and can cycle instruments to keep its power load below the generated power. [1] The mission's principal investigator, Alan Stern, estimates that the spacecraft has enough power to keep operating until the late 2030s. [28]

3.8 MMRTG Missions: (2011-Present (2021))

3.8.1 Curiosity

The Mars Science Laboratory (MSL) rover, named Curiosity (Figure 3.26), launched on 26 November 2011 and landed in Gale Crater on Mars on 5 August 2012, using guided entry and a sky crane system—new techniques that enhance NASA's capabilities for delivering payloads to the surface of Mars. In addition, it is the first NASA mission to use the Multi-Mission Radioisotope Thermoelectric Generator (MMRTG). MSL is a large, powerful science rover designed to study the layered rocks of Gale Crater on the Red Planet (Figure 3.27). Its primary objectives are to determine habitability (whether surface conditions are or were suitable to support life), characterize the planet's climate and geology, and help prepare for human exploration. Curiosity is the fourth rover the United States has sent to Mars and the largest, most capable rover ever sent to study a planet other than Earth. [1]

The Curiosity rover has a mass of 899 kg and is a six-wheeled rover that is about the size of a small sports utility vehicle: about 3 meters long (not including the reach of the arm), 2.7 meters wide, and 2.2 meters tall. Its arm reaches about

Figure 3.26 Curiosity took this self-portrait on 11 May 2016. Credit: NASA.

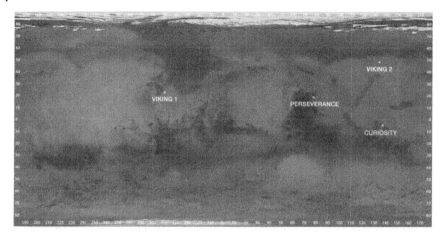

Figure 3.27 Mars map showing landing sites of Viking 1, Viking 2, Curiosity, and Perseverance. Credit: NASA.

2.2 meters. The Curiosity mission was enabled by its single MMRTG, which is based on the SNAP-19 RTGs flown on the two Viking landers and the Pioneer 10 and 11 spacecraft. Mission operators use excess heat from an MMRTG as a convenient and steady source of warmth to maintain proper operating temperatures of spacecraft electronics, instruments, or other components [29] on Curiosity. The MMRTG is the only RTG in production as of 2022.

The primary mission lasted one Mars year (about 23 Earth months) and used the Sample Analysis Mars (SAM) instrument suite, which was the most complex analytical chemistry laboratory delivered to another planet. Curiosity continued to operate in its extended mission in early 2022, collecting Martian soil samples and rock cores, and analyzing them for organic compounds and environmental conditions that could have supported microbial life now or in the past.

One MMRTG powers Curiosity and charges its two, 86 Ah Li-ion batteries. The MMRTG output was 114 W_e at the beginning of the surface mission using the 1957 W_t of its heat source. [30] The MMRTG was still healthy as of 2021, providing over 80 W_e of power nine years after landing during its continuing extended mission (Figure 3.28).

The DOE fueled Curiosity's RTG on 25 October 2008. Shortly after that, NASA chose to delay the launch for two years due to technical challenges in the completion of the rover. The Idaho National Laboratory maintained the fueled MMRTG in a storage configuration designed to minimize power degradation by lowering the hot-junction temperature of the thermoelectrics. Because of the age of the MMRTG at the time of launch, it was necessary to re-plan the surface missions with a lower power prediction of 110 W_e at landing. [30] The mission has been a great success, enabled by RTG power to support a long-duration surface mission with an ambitious traverse of the Martian surface.

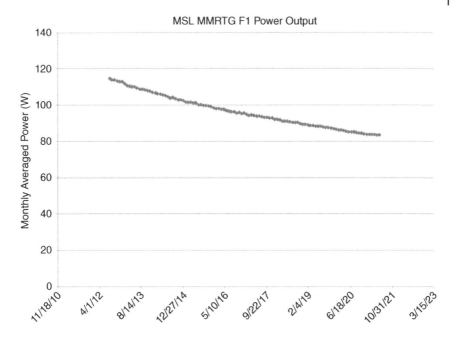

Figure 3.28 Curiosity Rover MMRTG monthly averaged power. Credit: Department of Energy/Idaho National Laboratory.

3.8.2 Perseverance

The Mars 2020 rover, named Perseverance, launched on 30 July 2020, and landed successfully in Jezero Crater on 18 February 2021. The Perseverance rover (Figure 3.29) is conducting geological assessments of its landing site on Mars, determining the habitability of the environment, searching for signs of ancient Martian life, and assessing natural resources and hazards for future human explorers. It is also collecting multiple rock and atmospheric samples for a potential return to Earth by a future mission. [1]

To keep mission costs and risks as low as possible, the Mars 2020 design was based on NASA's successful Mars Science Laboratory mission architecture, including its Curiosity rover and proven landing system. [1] Perseverance's central body and other major hardware, such as the cruise stage, descent stage, and aeroshell/heat shield, leveraged the success of the Curiosity rover and included many inherited components, such as the use of MMRTG and electrical power system. The car-sized Perseverance rover has similar dimensions to Curiosity: it's about 3 meters long (not including the reach of the arm), 2.7 meters wide, and 2.2 meters tall (10 feet long, 9 feet wide, and 7 feet tall). But at 1,025 kilograms, Perseverance is about 126 kilograms heavier than Curiosity. [31]

Figure 3.29 Artist's rendition of Perseverance rover, with MMRTG on the left top corner. [1], NASA.

Perseverance tested new technology, including an autopilot for avoiding hazards during landing called Terrain Relative Navigation (TRN) to allow the rover to land in more challenging terrain to benefit future robotic and human exploration of Mars. Perseverance also ferried a helicopter, called Ingenuity, the first aircraft with powered and controlled flight on another planet.

Like Curiosity, a single MMRTG powers Perseverance. At the time of landing, the MMRTG's thermal inventory was 1949 W_t, which resulted in 116 W_e of electrical power at the beginning of the surface mission (Figures 3.30, 3.31). The thermal power for conversion was 1949 W_t at time of landing. As of early 2022, the MMRTG has been on the surface for twelve months and is operating as expected, including the successful drilling and sealing of several rock cores.

As with Curiosity, the Perseverance mission was enabled by RTG power, which provides constant, reliable electrical and thermal power and environmental conditions day and night.

3.8.3 Dragonfly—Scheduled Future Mission

Dragonfly is in development as a mission to Titan under NASA's New Frontiers program, with a planned launch in 2027. [32] This mission intends to use an MMRTG to power a dual-quadcopter to fly from location to location on Titan. The Dragonfly design (Figure 3.32) has the potential to fly tens of kilometers in a single flight and make a flight each Titan day, or every 16 Earth days. Over its

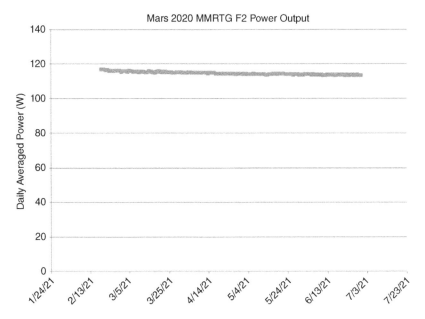

Figure 3.30 Perseverance Rover MMRTG Daily Averaged Power. Credit: Department of Energy/Idaho National Laboratory.

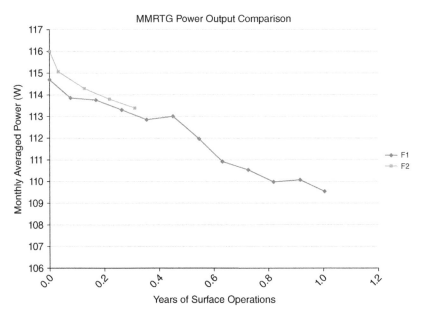

Figure 3.31 Perseverance MMRTG (F2) power output during surface operations, compared to Curiosity MMRTG (F1) for the first year. Credit: Department of Energy/Idaho National Laboratory.

Figure 3.32 Conceptual art of Dragonfly dual-quadcopter on Titan, with RTG enshrouded at the rear of the vehicle. [33], NASA.

two-year surface mission, the mission plans to traverse hundreds of kilometers. Along the way, the spacecraft will sample surface material and measure composition at each landing site, and monitor atmospheric and seismic conditions. [33] An RTG enabled the mission because solar power was insufficient at the bottom of Titan's hazy atmosphere; the spacecraft's need for electrical and thermal power during the long Titan nights is too great for a solar powered design. [33]

Dragonfly is designed for a predicted RTG output of 70 W_e at Titan using a single MMRTG. [34] The spacecraft team plans to perform flight operations, data transmission, and most science operations during the Titan day and will have plenty of time to recharge its Li-ion batteries during each Titan night.

3.9 Discussion of Flight Frequency

The history of United States space missions enabled by RTGs, summarized in Table 3.1, demonstrates decades of success in supporting many significant space exploration missions with RTG power. Figures 3.33 and 3.34 show the NASA and non-NASA radioisotope missions, respectively, categorized by destination.

By plotting these missions binned by decade in Figure 3.35, we can visualize the cadence of past, current, and upcoming RTG-powered missions.

Eight missions flew with RTGs in the 1960s, utilizing various options from the SNAP series of RTGs. Five missions were US Navy satellites in the Transit series,

Table 3.1 The United States' historic, current, and scheduled to be launched spaceflight missions powered by RTGs. The abbreviations in this table are defined in the section of this book entitled, List of Acronyms, Abbreviations, and Symbols.

Mission Name	Sponsor(s)	Type of RTG	Number of RTGs	Actual Average Power, W_e (per unit)[1]	RTG mass, kg (per unit)	# RHUs (1 W_{th} each)	Mission Type	Launch Date	Mission Status or End Date	RTG status
Transit 4-A	USN/APL	SNAP-3B	1	2.6[2]	2.1		Earth orbiter (navigation)	29 June 1961	1976	RTG operated for 15 years until Spacecraft shut down
Transit 4-B	USN/APL	SNAP-3B	1	3.1	2.1		Earth orbiter (navigation)	15 November 1961	August 1962	RTG operated for 9 years until last contact
Transit 5BN-1	USN/APL	SNAP-9A	1	25.2	12.3		Earth orbiter (navigation)	28 September 1963	June 1964	RTGs operated for 9 months until Spacecraft failure
Transit 5BN-2	USN/APL	SNAP-9A	1	26.8	12.3		Earth orbiter (navigation)	5 December 1963	1964	RTG operated for 6 years until last contact
Transit 5BN-3	USN/APL	SNAP-9A	1	-	12.3		Earth orbiter (navigation)	21 April 1964	20 April 1964	Mission aborted during launch
Nimbus-B	NASA	SNAP-19B	2	-	15.2		Earth orbiter (weather)	18 May 1968	18 May 1968	Mission aborted during launch
Nimbus III	NASA	SNAP-19B	2	28.2	15.2		Earth orbiter (weather)	14 April 1969	1971	RTGs operated for 2.5 years until Spacecraft shut down

(*Continued*)

Table 3.1 (Continued)

Mission Name	Sponsor(s)	Type of RTG	Number of RTGs	Actual Average Power, W_e (per unit)[1]	RTG mass, kg (per unit)	# RHUs (1 W_{th} each)	Mission Type	Launch Date	Mission Status or End Date	RTG status
Apollo 12	NASA	SNAP-27	1	73.6	19.7		Lunar lander (manned)	14 November 1969	September 1977	RTG operated for ~8 years until ALSEP was shut down
Apollo 13	NASA	SNAP-27	1	-	19.7		Lunar lander (manned)	11 April 1970	September 1977	Spacecraft failure prevented a successful lunar landing
Apollo 14	NASA	SNAP-27	1	72.5	19.7		Lunar lander (manned)	31 January 1971	September 1977	RTG operated for ~6.5 years until ALSEP was shut down
Apollo 15	NASA	SNAP-27	1	74.7	19.7		Lunar lander (manned)	26 July 1971	September 1977	RTG operated for ~6 years until ALSEP was shut down
Pioneer 10	NASA/ARC	SNAP-19	4	40.7	15.2	12	Interplanetary flyby	2 March 1972	January 2003 (last contact)	RTGs operated for > 31 years
Apollo 16	NASA	SNAP-27	1	70.9	19.7		Lunar lander (manned)	16 April 1972	September 1977	RTG operated for ~5.5 years until ALSEP was shut down
TRIAD	USN/APL	Transit-RTG	1	35.6	13.6		Earth orbiter (navigation)	2 September 1972	1982+	RTG operated for >10 years
Apollo 17	NASA	SNAP-27	1	75.4	19.7		Lunar lander (manned)	7 December 1972	September 1977	RTG operated for ~5 years until ALSEP was shut down

Pioneer 11	NASA/ ARC	SNAP-19	4	39.9	15.2	12	Interplanetary flyby	5 April 1973	November 1995 (last contact)	RTGs operated for > 22 years
Viking 1	NASA	SNAP-19	2	42.3	15.2		Mars lander	20 August 1975	November 1982	RTGs operated for 6 years until Lander was shut down
Viking 2	NASA	SNAP-19	2	43.1	15.2		Mars lander	9 September 1975	April 1980	RTG operated for 4 years until relay link was lost
LES 8	USAF / MIT	MHW-RTG	2	154	38		Earth orbiter	15 March 1976	June 2004	RTGs operated for 28 years until Spacecraft shut down
LES 9	Lincoln Laboratory USAF / MIT	MHW-RTG	2	154	38		Earth orbiter	15 March 1976	May 2020	RTGs operated for 44 years until Spacecraft shut down
Voyager 1	Lincoln Laboratory NASA/JPL	MHW-RTG	3	157	38	9	Interplanetary flyby, extrasolar	5 September 1977	Continuing	RTGs still operating
Voyager 2	NASA/JPL	MHW-RTG	3	159	38	9	Interplanetary flyby, extrasolar	20 August 1977	Continuing	RTGs still operating
Galileo	NASA/JPL	GPHS-RTG	2	288	56	120	Jupiter orbiter	18 October 1989	September 2003	RTGs operated for ~14 years until Spacecraft de-orbited

(*Continued*)

Table 3.1 (Continued)

Mission Name	Sponsor(s)	Type of RTG	Number of RTGs	Actual Average Power, W_e (per unit)[1]	RTG mass, kg (per unit)	# RHUs (1 W_{th} each)	Mission Type	Launch Date	Mission Status or End Date	RTG status
Ulysses	NASA/ESA	GPHS-RTG	1	289	56		Solar/space physics	6 October 1990	June 2009	RTG operated for more than 21 years until loss of contact
Cassini	NASA/JPL	GPHS-RTG	3	296	56	117	Saturn orbiter	15 October 1997	Sept 2017	RTGs operated for ~20 years until Spacecraft de-orbited
New Horizons	NASA/APL	GPHS-RTG	1	246	56		Pluto flyby	19 January 2006	Continuing	RTG still operating
Mars Science Laboratory (Curiosity)	NASA/JPL	MMRTG	1	115	45		Mars rover	26 November 2011	Continuing	RTG still operating
Mars 2020 (Perseverance)	NASA/JPL	MMRTG	1	118	45		Mars rover	30 July 2020	Continuing	RTG still operating
DragonFly	NASA/APL	MMRTG	1	Est. 70[3]	45		Titan multi-rotor vehicle	Expected 2027	Pre-Launch	F-3 Unit likely allocated to Dragonfly

[1] https://www.energy.gov/ne/articles/what-radioisotope-power-system

[2] 2.6 W was obtained from the APL Final Report on the Transit series of satellites, which was published in 1978 (SDO-1600 Revised). 2.8 W is regularly quoted in publications about this RTG. We are reporting 2.6 W because based on the available evidence, the APL report appears to be a better source.

[3] 70 W was a very conservative estimate published in 2018 of the power needed when the Dragonfly mission arrives on station 9 years after launch. The actual power is likely to be higher, as Figure 8.1 in Chapter 8 shows that 8.5 years after launch (0.8 years of cruise and 7.6 years on Mars) the MMRTG on Curiosity is producing 86 W.

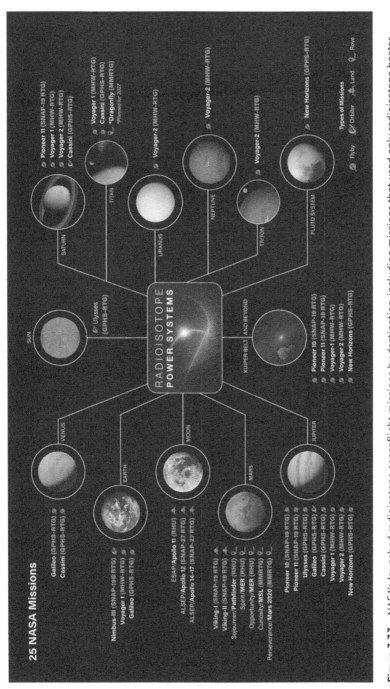

Figure 3.33 NASA's launched radioisotope spaceflight missions by destination, including missions that used only radioisotope heater units. Credit: NASA.

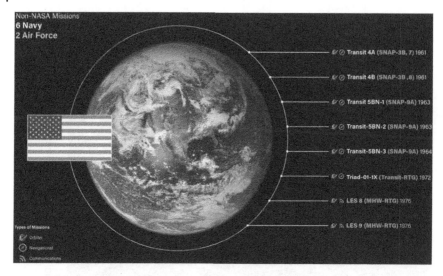

Figure 3.34 Non-NASA launched radioisotope spaceflight missions. Credit: NASA / Wikipedia Commons / Public domain.

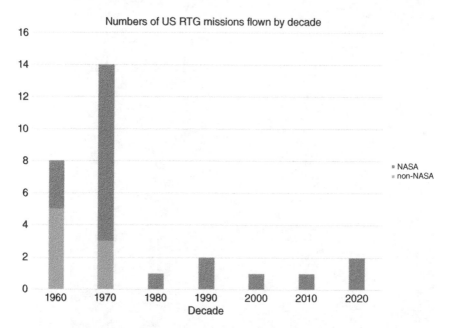

Figure 3.35 Past, current, and upcoming US RTG missions by decade.

performing primarily navigational research. The remaining three missions were two NASA missions conducting meteorological research and Apollo 12, the first of several Apollo missions using RTG power for long-term lunar surface experiments. Again, these missions used multiple options from the SNAP series of RTGs, demonstrating continuous improvements based on lessons from each successful mission.

The 1970s saw a rapid deployment of RTG-based NASA science missions. The development cycles for early RTGs were aggressive, innovating multiple SNAP generators and heat sources, including the intact re-entry heat source (IRHS) and a new heat source for the MHW-RTG. As the RTGs became more capable and reliable, they were used to power more ambitious science missions: Pioneer, Viking, and Voyager.

A decade after the MHW-RTG powered the Voyager missions, the GPHS-RTG had its first flight, followed by three more flights over the next two decades. Then the MMRTG had its first two flights, with another (Dragonfly) expected in the 2020s. NASA flew fewer RTG missions across these decades, as RTGs completed their progression from an experimental technology to a mature technology used on complex and high-budget missions with a slower cadence of flights.

Figure 3.36 shows the total RTG power of US missions flown in each decade. While the 1960s saw seven US RTG missions flown, the total power across these

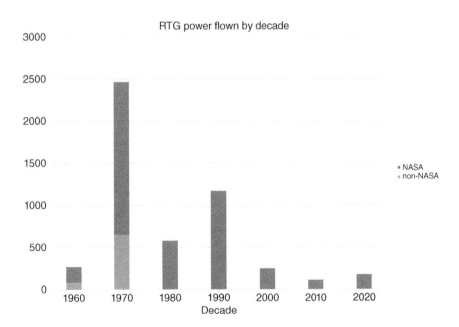

Figure 3.36 RTG power in space from past, current, and upcoming US missions by decade.

missions was low. The RTG units were relatively low-power systems and often accompanied by solar power. In the 1970s, in contrast, the combined power provided to numerous missions from the higher power MHW-RTG for a total power across the decade was ~2.5 kW$_e$.

The 1990s, despite only flying two RTG missions, stand out in second place. The available RTG in that decade was the high-power GPHS-RTG, providing ~300 W$_e$ per RTG at BOM. Three flew on Cassini alone and demonstrated the science-producing capability of high-power RTGs for deep space science.

Flights in recent decades have seen lower power levels on RTG missions. New Horizons launched carrying a spare Cassini GPHS-RTG, and after that the smaller, lighter, and lower-power MMRTG has been the only available RTG. This RTG enabled roving Mars missions and the upcoming Titan lander, because it could operate in a planetary atmosphere. Mission designers have proposed missions requiring nearly one kW$_e$, as the Cassini mission did, but those will not fly before higher-power RTGs that NASA and the DOE are developing become available.

The number of RTG units launched in Figure 3.37 shows the 1970s standing out more starkly, with most of those missions launching multiple units—the US launched 28 units on 14 missions. Compared to the 1960s missions, which often

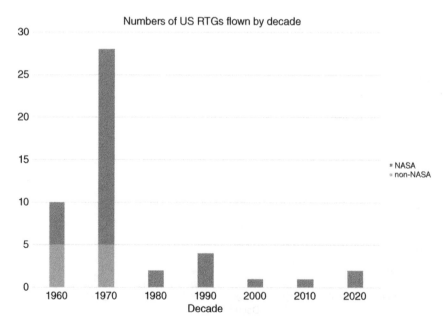

Figure 3.37 The number of RTG units launched by past, current, and upcoming US missions by decade.

used RTGs as auxiliary power sources, the missions in the 1970s relied on RTG power as their sole power source. As early RTGs produced less power relative to later RTGs, multiple units were frequently required to meet a mission's power needs. For instance, the Pioneer and Viking missions were significant contributors to the number of units flown, as they utilized 12 total SNAP-19 RTGs.

The US Department of Energy supports the anticipated future cadence of RTG missions by an ongoing, constant rate of production of plutonium dioxide sufficient to fuel 25-37 GPHS modules per decade. This could fuel 3 to 4 MMRTGs or 1 to 2 of an updated version of the GPHS-RTG per decade. The DOE is also taking steps to enable increasing production further.

3.10   Summary of US Missions Enabled by RTGs

RTGs have enabled many monumental and historical space missions for the USA for over 60 years, bringing back to Earth wonders and invaluable science returns that have answered many questions for humanity about our home world, the solar system, and interstellar space.

Pioneer 10 was the first spacecraft to visit the giant planets and the first human-made object to leave the solar system. Pioneer 11 was the first mission to explore Saturn. Viking 1 was the first mission to successfully land on Mars, finding the elements essential to life on Earth: carbon, nitrogen, hydrogen, oxygen, and phosphorus. With enduring power from RTGs, both Viking 1 and 2 operated far beyond their design lifetime of 90 days after landing on Mars. Galileo was the first spacecraft to orbit Jupiter, where it studied the planet and its rings for almost eight years. Ulysses was the first to survey the space environment above and below the Sun's poles. Cassini was the first spacecraft to orbit Saturn and closely explore many of its moons. It also carried the Huygens probe, which became the first to land on the surface of Titan, the first-ever landing in the outer solar system.

Those missions, Voyager 1 and 2, that left Earth in 1977 are still going strong, venturing to the edge of the solar system and beyond to broaden our understanding of space more than ever. New Horizons was the first spacecraft to observe Pluto and its system and is now heading to explore Kuiper Belt Objects. Both the Curiosity and Perseverance Mars rovers are exploring the Red Planet, looking for signs of ancient life. Dragonfly is on track to be the second competitively selected—rather than directed—NASA mission powered by an RTG, with unique, exciting science returns in store. The common theme across these missions is their long-lived power source, the RTG, which enabled some of humankind's greatest discoveries and long-duration space explorations! The future of space exploration remains bright and exciting.

References

1 https://rps.nasa.gov/missions/ (accessed 30 November 2022).
2 Schmidt, G.R., Sutliff, T.J., and Dudzinski, L.A. (2009). Radioisotope Power: A Key Technology for Deep Space Exploration. *IAC-09-C4.7.-C3.5.4.*
3 Bennett, G.L. and Skabek, E.A. (1996). *Power Performance of US Space Radioisotope Thermoelectric Generators.* IEEE.
4 Bennett, G.L., Lombardo, J.L., and Rock, B.L. (1984). US radioisotope thermoelectric generators in space, the nuclear engineer. 5 (2).
5 https://rps.nasa.gov/about-rps/safety-and-reliability/ (accessed 30 November 2022).
6 https://www.smithsonianmag.com/science-nature/day-nimbus weather-satellie-180961686/ (accessed 30 November 2022).
7 Fihelly, A.W. and Baxter, C.F. (1970). The SNAP-19 radioisotopic thermoelectric generator experiment. *IEEE Int. Geosci. Remote Sens. Symp.*
8 Bennett, G. (2008). Mission interplanetary: Using radioisotope power to explore the solar system. *Energy Convers. Manag.* 49 (3): 382–392.
9 Anderson, J.D., Laing, P.A., Lau, E.L., Liu, A.S., et al. (2002). Study of the anomalous acceleration of Pioneer 10 and 11. *Physical Review D.* 65 (8): 082004.
10 https://www.jpl.nasa.gov/missions/viking-1 (accessed 30 November 2022).
11 Barnett, W. (1989). Thermoelectric Alloys and Devices for Radioisotope Space Power Systems: State-of-the Art and Current Developments.
12 https://www.hq.nasa.gov/alsj/HamishALSEP.html (accessed 30 November 2022).
13 Ward, W.W. Developing, Testing, and Operating Lincoln Experimental Satellites 8 and 9. https://apps.dtic.mil/sti/citations/ADA069095 (accessed 30 November 2022).
14 https://news.mit.edu/2020/lincoln-laboratory-decommissions-lincoln-experimental- satellite-9-0527 (accessed 30 November 2022).
15 https://voyager.jpl.nasa.gov/mission/spacecraft/ (accessed 30 November 2022).
16 Kohlhase, E.C. and Penzo, P.A. (1977). Voyager Mission Description. *Space Sci. Rev.* 21: 77–101.
17 https://www.jpl.nasa.gov/news/a-new-plan-for-keeping-nasas-oldest-explorers-going (accessed 30 November 2022).
18 https://www.jpl.nasa.gov/missions/galileo (accessed 30 November 2022).
19 Galileo Orbiter Engineering Team End of Prime Mission, Final Performance and Trend Analysis Report. *IOM: GLL-OET-98-981-WJG.* (1998).
20 https://www.jpl.nasa.gov/missions/ulysses (accessed 30 November 2022).
21 Bennett, G.L., Hemler, R.J., Schock, A. (1994). Development and Use of the Galileo and Ulysses Power Sources, 45th Congress of the International Astronautical Federation.
22 https://solarsystem.nasa.gov/missions/ulysses/in-depth/ (accessed 30 November 2022).

23 Cassini Quick Facts. https://solarsystem.nasa.gov/missions/cassini/mission/quick-facts/ (accessed 30 November 2022).

24 Grandidier, J., Woerner, D.F., and Burke, T.A. (2018). Cassini Power During the 20-Year Mission and Until the Final Plunge into Saturn. *J. Spacecr. Rockets* 55 (6).

25 Ottman, G.K. and Hersman, C.B. (2006). The pluto-new horizons rtg and power system early mission performance. Poster presented at the 4th International Energy Conversion Engineering Conference and Exhibit, American Institute of Aeronautics and Astronautics.

26 NASA Space Science Data Coordinated Archive–Cassini. https://nssdc.gsfc.nasa.gov/nmc/spacecraft/display.action?id=1997-061A (accessed 30 November 2022).

27 http://pluto.jhuapl.edu/Mission/Spacecraft/Systems-and-Components.php (accessed 30 November 2022).

28 https://spacenews.com/new-horizons-planning-additional-extended-missions/ (accessed 30 November 2022).

29 https://rps.nasa.gov/power-and-thermal-systems/power-systems/ (accessed 30 November 2022).

30 Woerner, D., Moreno, V., Jones, L. et al. (2013). *The Mars Science Laboratory (MSL) MMRTG In-Flight: A Power Update.* NETS.

31 https://mars.nasa.gov/files/mars2020/Mars2020_Fact_Sheet.pdf (accessed 30 November 2022).

32 Lorenz, R.D., MacKenzie, S.M., Neish, C.D., Gall, A.L. et al. (2021). Selection and characteristics of the dragonfly landing site near Selk Crater, Titan. *Planet. Sci.* 2 (24): 1–13.

33 https://dragonfly.jhuapl.edu/ (accessed 30 November 2022).

34 Lorenz, R.D., Turtle, E.P., Barnes, J.W., Trainer, M.G., et al. (2018). Dragonfly: a rotorcraft lander concept for scientific exploration at Titan. *Johns Hopkins APL Tech. Dig.* 34 (3): 1–14.

4

Nuclear Systems Used for Space Exploration by Other Countries

Christofer E. Whiting

University of Dayton Research Institute, Dayton, Ohio

While the United States is undoubtedly the world leader in spaceborne RTGs, a few other countries have also attempted to develop this capability. The European Space Agency is probably the most well-known because they have been very open about developing their americium-241 RTG. [1] China and the Soviet Union are also known to have launched radioisotope power systems into space, but detailed discussions of their space programs are not in the open literature/media. This absence of information makes it very difficult to find authoritative references that can discuss, or even confirm, the use of Soviet/Russian or Chinese RTGs in space. Thus, making it hard to appreciate the RTG capabilities held by these other countries.

Here we will review the verified capabilities of other countries with authoritative media or literature sources.

4.1 Soviet Union[1]

The Soviet Union still holds the record for most nuclear space mission launches in the world. [2,3] However, most of these missions were powered by space reactors that only operated in Earth orbit for a few weeks or months. Between 1969 and 1988, the Soviet Union launched thirty-five nuclear reactors that achieved stable orbit. The vast majority of these reactors were of the RORSAT design, which used

---

1 Nearly all of the nuclear launches occurred under the Soviet Union flag, so the term Soviet Union or Soviet will be applied to all blanket discussions of this program. The term Russian Federation or Russia will be used in the few instances when a nuclear mission occurred under the Russian Federation flag.

The Technology of Discovery: Radioisotope Thermoelectric Generators and Thermoelectric Technologies for Space Exploration, First Edition. Edited by David Friedrich Woerner.

a highly enriched uranium-based fission core to produce heat and thermoelectric couples to convert the heat into electricity. [3] Two of the Soviet reactors launched into space were TOPAZ reactors that used a thermionic power conversion system to harness heat from the reactor to generate electricity. [3] Thermionic converters have been of particular interest for both the Soviet Union and the US because of the promise of much higher conversion efficiencies. Unfortunately, the extremely high temperatures required to operate thermionic converters creates a number of major engineering challenges. The US investigated the use of thermionic conversion for the SNAP-13[2] but the timeframe and budget available were not sufficient to resolve the technical challenges with SNAP-13.

In 1988, cutbacks to the Soviet research budgets ended the TOPAZ-2, a thermionic-based advanced nuclear reactor program. Seeing a unique opportunity to take advantage of the state-of-the-art Soviet understanding of both space nuclear reactors and thermionic technology, an international collaboration between the Russian Federation, USA, UK, and France explored using the twenty-six TOPAZ-2 systems built by the Soviet Union for ground testing and potential flight applications. Unfortunately, this collaborative program could never meet its total objectives, which relegated the TOPAZ-2 systems to ground testing. [3]

The Soviet space reactor program history is rich enough to be the subject of several book chapters. Unfortunately, a deep dive into this history would be off-topic. Interested readers can find an authoritative overview of the Soviet space reactor program in References 3 and 4.

The Soviets used radioisotope power systems on a handful of space missions. Cosmos[3,4] 84 and 90 are believed to have flown Orion-1[5] RTGs in 1965, only a few years after the US launched its first SNAP-3A. [3] Orion was a polonium-210 based RTG that used a silicon semiconductor-based thermoelectric converter with about a 20 W_e power output. [4] These RTGs were believed to be useful for ~3000 h, or ~4 months. This short lifetime was because of the short (138 day) half-life of

---

2 SNAP-13 was developed for the NASA Surveyor program, which landed a series of robotic spacecraft on the moon. SNAP-13 never evolved past the initial design testing phase.

3 Due to the lack of communication from the Soviet Union about their space program, the Soviet mission designations were often not known. Beginning in 1962, NASA assigned the name Cosmos along with a sequential numerical designation to Soviet spacecraft, which remained in Earth orbit, regardless of whether that was their intended final destination.

4 There is disagreement in the literature regarding whether the designation should be Cosmos, or the Greek Kosmos. Since the designation is an arbitrary NASA designation, we are using Cosmos, which is the spelling used by NASA's primary sources.

5 One IAEA report identifies Orion-1 as the RTG that flew on Cosmos 84 and Orion-2 as the RTG that flew on Cosmos 90. [3] This report does not provide references supporting these designations. Bennett also provides an authoritative reference on the Soviet space reactor program, and he calls both RTGs Orion-1. The Bennett report does provide a Soviet based reference for his designation, so we are inclined to believe that the Bennett designation is more accurate.

Figure 4.1 Model of the Lunokhod 1 rover in the Museum of Cosmonautics (Moscow). Petar Milošević / Wikipedia Commons / CC BY-SA 3.0.

the radioisotope. See Chapter 3 for more details on ^{210}Po and its use in the US radioisotope program.

The Soviets also used radioisotopes in a handful of lunar missions. For example, during the height of the space race, the Soviet Luna program targeted several scientific missions at the moon. The most ambitious was the Lunokhod series of robotic rovers (Figure 4.1). These rovers carried an array of photographic and scientific equipment and were powered by batteries that were recharged using a small solar array. However, the rover would close up during the long lunar nights, and a ^{210}Po radioisotope heater unit would keep the internal components warm. [3] Reports suggest the amount of heat produced by the RHU was between 900 W [4] and 800 W. [5] Given the short half-life of ^{210}Po, the thermal power of the RHU would decay from 900 to 800 W in a mere 20 days. Therefore, we find these references are likely not disagreeing but instead referring to the thermal inventory of the RHU at different points in time (e.g., fueling, launch, or arrival on the moon).

The first lunar rover was launched by the Soviet Union in February 1969.[6] This mission exploded less than a minute after leaving the Earth's surface. Because this

---

6 Soviet internal designation – Ye-8 number 201. [6]

mission never achieved orbit, it is technically not a "space mission" and it was never given a Cosmos number by NASA. The convention used in this book, however, is that a nuclear launch is one that lift offs, and this mission meets that standard. Unfortunately, the level of secrecy around the Soviet space program meant that the rest of the world did not know about this launch, or the status of the nuclear payload, until years later. Despite an intensive search, the ^{210}Po heat source was never reported as found. [6]

Lunokhod I was the first wheeled vehicle on another world, and was one of the most successful robotic missions of the early space age, with a mission duration of 322 days. [6] Lunokhod II was also a resounding success and set a record of 42 kilometers (26 miles) for the longest distance traveled by a vehicle on another celestial body. Lunokhod II would hold onto that record for 42 years before NASA's Martian rover Opportunity finally overtook it in 2015. These firsts in space exploration could not have been possible without radioisotope heat.

There is speculation that Cosmos 300 and Cosmos 305 may have also included payloads with radioisotope heater units. [3,7] Bennett bases this speculation on the notion that these missions may have been Lunokhod rovers. Unfortunately, both of these missions experienced upper-stage malfunctions that prevented the spacecraft from leaving Earth's orbit. According to NASA's website, both spacecraft were later determined to be lunar sample return missions. [8,9] Unlike a rover, a lunar sample return would not require a radioisotope heat source to do its job. Plus, the Soviet Union launched several other lunar sample return missions that are not cited as having radioisotope heater units. Given that the basis of Bennett's speculation does not appear to be accurate, we are inclined to believe that Cosmos 300 and 305 were not nuclear launches.

After the collapse of the Soviet Union, the Russian Federation launched one radioisotope mission. Mars 96 (aka Mars-8) was an ambitious mission with an orbiter, two small landers, and two separate penetrators. [6] The two ground stations and two penetrator assemblies on Mars 96 possessed a combined 200 grams of ^{238}Pu as a radioisotope heat source that was split between several RHUs and RTGs that produced 8.5 W of heat and 0.2 W of electricity, respectively. The RHU and RTG designs were named Angel. [4,5,10] We speculate that the purpose of these heat sources was for the thermal management of sensitive electronics during the cold Martian nights, but this is not expressly stated in any authoritative source.[7] Unfortunately, a failure in the fourth stage burn caused the probe assembly to reenter the Earth's atmosphere, and the mission was lost.

---

7 Nearly every Martian surface mission has needed some method of staying warm, and using electrical heat to stay warm would be a major drain on the mission power resources. Many of the US's surface missions to Mars have used radioisotope heat for this reason.

4.2   China

Recently, China has joined the very short list of countries that have deployed radi-oisotope power in space. In December 2013, China launched the Chang'E-3 lunar mission. Chang'E-3 deployed a soft lander onto the surface of the moon. About a week later, the lander deployed the lunar rover "Yutu." Both the lander and rover possess ^{238}Pu RHUs to heat the robotic systems during the long lunar night. [6,11] Reports do not discuss the size of the Chang'E-3 RHUs. It is likely that this mission either uses RHUs similar to Chang'E-4 (see below) or Angel RHU technology.

Unfortunately, Yutu lost the ability to move after only about a month when it experienced a "mechanical control abnormality." [12] Yutu's mission was adjusted to become a stationary science platform, and it operated for a total of 31 months (July 2016). The Chang'E-3 lander continues to function and is the longest-lasting spacecraft to operate on the lunar surface.

In January 2019, Chang'E-4 became the first mission to accomplish a landing on the moon's far side. Both the lander and Yutu-2 rover have RHUs to keep the spacecraft warm during the long lunar night. Three RHUs have been reported on the lander. One produces 120 W of heat, while the other two produce 5 W each. [13] The lander, however, also has a ^{238}Pu RTG[8] as an auxiliary 2.5 W electrical power source [13-17], making China the second country to provide electricity beyond Earth's orbit using radioisotopes. The RTG also produces 120 W of heat [13], which suggests that the RTG is based on the 120 W RHU design. However, there is no discussion on the size of the RHU on Yutu-2. One report, however, measures the radiation dose on the moon before and after the Yutu-2 disembarked from the lander. [14] The difference in radiation dose rates after the Yutu-2 left is too small for a 120 W ^{238}Pu heat source, suggesting instead that the Yutu-2 RHU is of the 5 W variety.

Other notable firsts for Chang'E-4 include the Lunar Micro Ecosystem experiment. [18] This sealed cylinder attempted to create a functional biosphere on the moon using various plant seeds, yeast, and fruit fly eggs. Several plant seeds sprouted, but a failure in the temperature regulation caused the experiment to fail during the very cold lunar night. [19,20] The addition of the RTG also allowed the Chang'E-4 lander to measure nighttime lunar soil temperatures on the far side of the moon.[9] Temperatures as low as -196 °C (-320.8 °F) were observed. In December

---

8 From the references reviewed here it is clear the Chang'E-4 lander has an RTG, but the writing is ambiguous regarding whether the Yutu-2 also has an RTG. In our opinion, the context of most references implies that the lander, and only the lander, has a single RTG.

9 This is a scientific first because the moon is tidally locked, meaning the same side is always facing the Earth. All lunar surface exploration prior to Chang'E-4 was on the near side of the moon. Infrared radiation from the warm Earth will influence temperature measurements on the near side.

2019, Yutu-2 became the longest-lasting rover on the moon. In February 2020, Yutu-2 was the first mission to image a lunar ejecta sequence and directly analyze its internal architecture. [21] Both the lander and the Yutu-2 still appear to be operational.

There has been some, perhaps unfair, rumor and speculation in the space exploration community that Russia supplied the RHUs and RTGs used by China. China has not been very open about its development of RTG technology, but according to some authoritative sources, one significant advance of Chang'E-4 was the indigenous development of RTG technology. [17] These references do not discuss or make inferences regarding the RHU technology on Chang'E-3.

There is some disagreement in the literature regarding whether Chang'E-3 also had an RTG. Most references are not close enough to the Chang'E-3 mission to be authoritative.[10] Other authoritative sources might have stated Chang'E-3 was going to have an RTG, but in our research, these sources were written before the mission was finalized and launched. Authoritative sources only talk about RHUs on Chang'E-3 and Yutu [6], including sources from within the Beijing Institute of Spacecraft Engineering. [11,15][11] This includes a book by Chen *et al.*, which states, "The power system of Chang'E-4 inherits the power system of Chang'E-3, but uses an RTG as an auxiliary power source...". [15] Given the authoritative nature of these sources, it would seem clear that Chang'E-3 did not possess an RTG.

References

1 Ambrosi, R., Williams, H., Samara-Ratna, P. et al. (2013). Americium-241 radioisotope thermoelectric generator development for space Applications. *Proceeding of International Nuclear Atlantic Conference – INAC 2013*, Recife, PE (November 2013).

2 Summerer, L., Stephenson, K., Safa, F., and Kminek, G. (2012). *Nuclear Power Sources for Space Applications – A Key Enabling Technology*. Manchester, UK: *Space NPS Applications.*

---

10 Some of these sources incorrectly indicate that Chang'E-3 used a GPHS-RTG, which is a US design that definitely has not been deployed on any Chinese missions. GPHS-RTG is also taller than the Chang'E-4 lander, so if it were on any of the Chang'E missions, it would be obvious from the pictures.

11 The Beijing Institute of Spacecraft Engineering is subordinate to the government owned China Aerospace Science and Technology Corporation, which is the prime contractor for the Chinese space program.

3 Bennett, G.L. (1989). A look at the Soviet Space Nuclear Power Program. *Proceedings of the Intersociety Energy Conversion Engineering Conference (IECEC-89),* Washington, D.C. https://doi.org/10.1109/IECEC.1989.74620.

4 Stanculescu, A. (ed.) (2005). *The Role of Nuclear Power and Nuclear Propulsion in the Peaceful Exploration of Space,". STI/PUB/1197.* Vienna, Austria: International Atomic Energy Agency. ISBN: 92-0-107404-2.

5 Chmielewski, A.B., Borshchevsky, A., Lange, R., and Cook, B. (1994). A survey of current russian RTG capabilities. *Proceedings of the 29th Intersociety Energy Conversion Engineering Conference*, Monterey, CA. American Institute of Aeronautics and Astronautics, Washington, D.C.

6 Siddiqi, A.A. (2018). *Beyond Earth," NASA SP-2018-4041.* Washington, D.C: National Aeronautics and Space Exploration, Office of Communications. ISBN: 9781626830424.

7 Launius, R.D. (2015). Powering Space Exploration: Space Nuclear Propulsion, Public Perceptions, and Planetary Probes. www.academia.edu/11989657/Powering_Space_Exploration_RTGs_Nuclear_Reactors_and_Outer_P lanetary_Probes (accessed 30 November 2021).

8 Anon. (2022). NASA Space Science Data Coordinated Archive: Cosmos 300. *NSSDCA/COSPAR ID: 1969-080A.* nssdc.gsfc.nasa.gov/nmc/spacecraft/display.action?id=1969-080A (accessed 30 November 2021).

9 Anon. (2022). NASA Space Science Data Coordinated Archive: Cosmos 305. *NSSDCA/COSPAR ID: 1969-092A.* nssdc.gsfc.nasa.gov/nmc/spacecraft/display.action?id=1969-092A (accessed 30 November 2021).

10 Summerer, L. (2006). Technical Aspects of Space Nuclear Power Sources – VII. Radioisotope Heater Units. *ACT-RPT-2327-RHU.* International Atomic Energy Agency, Vienna, Austria.

11 Sun, Z., Jia, Y., and Zhang, H. (2013). Technological advancements and promotion roles of chang'E-3 lunar probe mission. *Sci. China Technol. Sci.* 56 (11): 2702–2708. https://doi.org/10.1007/s11431-013-5377-0.

12 Jones, A. (2020). China's Chang'E-3 Lunar Lander still going strong after 7 years on the Moon. *Space.com* (23 September) www.space.com/china-change-3-moon-lander-lasts-7-years (accessed 30 November 2021).

13 Hou, D., Zhang, S., Yu, J. et al. (2020). Removing the dose background from radioactive sources from active dose rate measurements in the lunar Lander Neutron & Dosimetry (LND) experiment on Chang'E 4. *J. Instrum.* 15: P01032. https://doi.org/10.1088/1748-0221/15/01/P01032.

14 Zhang, S., Wimmer-Schweingruber, R.F., Yu, J. et al. (2020). First Measurements of the Radiation Dose on the Lunar Surface. *Sci. Adv.* 6 (39): 1–6. https://doi.org/10.1126/sciadv.aaz1334.

15 Chen, Q., Liu, Z., Zhang, X., and Zhu, L. (2020). *Spacecraft Power System Technologies*. Singapore: Beijing Institute of Technology Press and Springer Nature Singapore. ISBN: 978-981-15-4838-3.

16 Li, C., Zuo, W., Wen, W. et al. (2021). Overview of the Chang'e-4 Mission: Opening the Frontier of Scientific Exploration of the Lunar Far Side. *Space Sci. Rev.* 217 (35): 1–32. https://doi.org/10.1007/s11214-021-00793-z.

17 Wu, W., Yu, D., Wang, C. et al. (2020). Technological Breakthroughs and Scientific Progress of the Chang'e-4 Mission. *Sci. China Inf. Sci.* 63 (200201): 1–14.

18 Huan, J., Xiao, Z., Flahaut, J. et al. (2018). Geological Characteristics of Chang'E-4 Landing Site. In: *Proceedings of the 49th Lunar and Planetary Science Conference (LPSC)*, 2083. TX: The Woodlands.

19 Liu Jia. (2019). Moon Sees First Cotton-Seed Sprout. xinhuanet.com/english/2019-01/15/c_137745432.htm or www.spacedaily.com/reports/Moon_sees_first_cotton_seed_sprout_999.html (accessed 30 November 2021).

20 Gough, E. (2019). We just got new information on the first-ever plant sprouted by China on the Moon. *Sciencealert.* sciencealert.com/china-s-lander-successfully-grew-some-cotton-plants-on-the-moon (accessed 30 November 2021).

21 Li, C., Su, Y., Pettinelli, E. et al. (2020). The Moon's farside shallow subsurface structure unveiled by Chang'E-4 Lunar Penetrating Radar. *Sci. Adv.* 6 (9): 1–8. https://doi.org/10.1126/sciadv.aay6898.

5

Nuclear Physics, Radioisotope Fuels, and Protective Components

Michael B.R. Smith[a], Emory D. Collins[a], David W. DePaoli[a],
Nidia C. Gallego[a], Lawrence H. Heilbronn[b], Chris L. Jensen[a],
Kaara K. Patton[a], Glenn R. Romanoski[a], George B. Ulrich[a],
Robert M. Wham[a], and Christofer E. Whiting[c]

[a]*Oak Ridge National Laboratory, Oak Ridge, Tennessee*
[b]*University of Tennessee, Knoxville, Tennessee*
[c]*University of Dayton Research Institute, Dayton, Ohio*

5.1 Introduction

The radioisotope thermoelectric generator (RTG) represents a unique device requiring a confluence of disciplines, to include aerospace, materials science, and nuclear engineering, on an extraordinary level. While the heart of an RTG is a deceptively simple-looking mass of radioisotope material, the methods to produce this material require a thorough understanding of some of the most fundamental yet complex phenomena in our universe. To explain how isotopes are created, processed, and used in an RTG, a comprehensive description must be provided. Although this chapter is not a complete discourse on nuclear physics, historic RTG fuel selections, and current RTG fuel and component fabrication, it does provide introductions to each of these topics.

This chapter provides foundational instruction on the nuclear and atomic knowledge relevant to radioisotope fuel behavior, RTG performance, and radiological concerns. To ensure that the reader gains a clear understanding of these concepts, familiarity with the atom, radioactivity, particle emission, nuclear reactions, and general nuclear terminology is required. This foundation will prepare the reader for further discussion regarding RTG fuel and component fabrication.

This chapter provides the reader with context to explain why certain isotopes are preferred in different mission architectures in relation to their fundamental nuclear

The Technology of Discovery: Radioisotope Thermoelectric Generators and Thermoelectric Technologies for Space Exploration, First Edition. Edited by David Friedrich Woerner.
© 2023 John Wiley & Sons, Inc. Published 2023 by John Wiley & Sons, Inc.

characteristics. An understanding of these characteristics is based on the foundation presented in the nuclear physics discussion, providing a platform for analysis of current plutonium oxide (PuO_2) production and protective components.

Sections on fuel fabrication and protective components for RTGs provide the reader with a macro-scale view of RTG fuels, as well as a discussion of the processes, facilities, and expertise required to fabricate, handle, and ship these complex materials. Furthermore, the contemporary fuel form of PuO_2 is presented, along with historic and current methods of its safe encapsulation.

References for further reading on each topic are provided throughout the chapter, and the reader is encouraged to investigate them at their convenience.

5.2   Introduction to Nuclear Physics

5.2.1   The Atom

The fundamental building blocks of matter are *atoms*, which consist primarily of subatomic particles called *neutrons*, *protons*, and *electrons*. While the known particle inventory of the standard model [1] shows many different particles of varying mass, charge, and matter/anti-matter states, this chapter is deliberately limited to discussion of these protons, neutrons, and electrons. A high-level presentation of these particles' characteristics can be found in Table 5.1.

These subatomic particles (when bound in an atom) can generally be categorized by their rest mass, charge, and region within the atom where they reside.

Table 5.1   High-level particle characteristics. Data referenced from the National Institute of Standards and Technology[1]

Name	Symbol	Charge	Mass (kg)	Mass (MeV)[a]
Proton	p^+	Positive	1.673×10^{-27}	938.272
Neutron	n^0	Neutral	1.675×10^{-27}	939.565
Electron	e^-	Negative	9.109×10^{-31}	0.511

[a] While discussion of mass thus far has used the common SI base-unit of kilograms (kg), many nuclear disciplines use the concept of the *electron volt* (eV) to describe the energy and/or mass of a given particle, ion, or atom. An electron volt (eV) is the amount of kinetic energy imparted to a single electron when exposed to a 1.0 V electric field. $1\ eV = 1.602 \times 10^{-18}$ J. Furthermore, many of the particle masses and kinetic energies relevant to RTG fuels will be on the order of 10^6 eV – or MeV. Further discussion of particle energies in this work will use MeV as the base unit, where $1\ MeV = 1.6 \times 10^{-13}$ J.

1 https://www.nist.gov/pml/fundamental-physical-constants

Protons and neutrons are similar in mass, but they carry different charges: protons carry a positive charge, and neutrons carry no charge. Also, protons and neutrons are approximately 1,800 times more massive than electrons, which carry a negative charge.

Protons and neutrons reside in the central region of the atom, the *nucleus*, which is held together by a force called the nuclear strong force. Electrons are bound in the outer-most regions of the atom in discrete groupings called *orbitals* or *clouds* by an attractive force to the protons in the nucleus called the *electromagnetic* (EM) or *Coulombic* force. [2,3] An illustration of a helium atom (not to scale) is provided in Figure 5.1 to show a notional layout, including the particle region, and the relative size of an atom.

When communicating the particle inventory of an atom in print (i.e., protons, neutrons, and electrons), a standard notation is used. Elements are represented by lettered symbols such as H, Pb, or W, which are familiar to anyone acquainted with the periodic table of elements. The number of protons in an atom is called the *atomic number* and is denoted by the letter Z in a leading subscript of the elemental symbol, whereas the neutron number N is denoted by a trailing subscript. In a neutrally charged atom, Z also represents the number of electrons. The mass number A is denoted by a leading superscript and represents the number of protons Z plus the number of neutrons N (A does not include electrons).

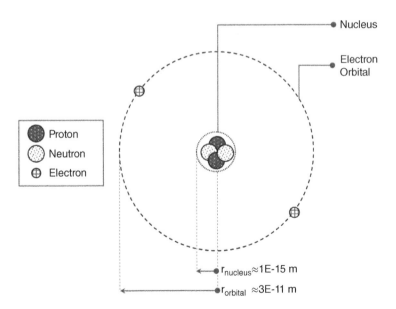

Figure 5.1 Illustration of a helium atom's fundamental particles, regions, and approximate dimensions.

While complete notation of a given atom may include the atomic number Z, the mass number A, and the neutron number N (e.g., $^A_Z X_N$), it is common practice to exclude Z and N and refer to a given isotope only by the elemental symbol and A, such as ^{238}Pu, ^{241}Am, or ^{90}Sr. Atoms with the same number of protons Z but with a different number of neutrons, $N = A-Z$, are called *isotopes* of the same element, as in ^{236}Pu, ^{238}Pu, and ^{239}Pu. Further mention of specific isotopes in this chapter will follow this common notation.

The *rest mass* of a particle is simply the mass of a given particle when it is not in motion and not bound in a nucleus—a largely hypothetical scenario. A proton and a neutron separated by distances in excess of ~10^{-15} m (the distance of influence of the nuclear strong force [3]) will have a distinct mass. However, if these two particles are bound in a nucleus ($<$~10^{-15} m apart), then the overpowering influences of the nuclear strong force will manifest, and a portion of the particle's original mass is converted into what is known as *binding energy*. This binding energy is what drives the energetics of many nuclear decays and reactions and is related to Einstein's famous mass-energy equivalence equation, $E = mc^2$. [4]

A helium nucleus serves as a working example: when two neutrons and two protons are brought together to form a nucleus, their experimentally measured mass in a bound state will be less than the original sum of their separated constituent parts. This difference in mass is referred to as the *mass defect* and can be used to calculate the nuclear binding energy of the atom using mass–energy equivalence. The *binding energy* is the amount of energy released when forming a nucleus, or the amount of energy required to separate that same nucleus.

The binding energy of a helium nucleus is 4.527×10^{-12} Joules (28.295 MeV c^{-2}). This may not seem like a large number, but when one considers that a single gram of helium contains ~10^{23} atoms, the potential nuclear energy that can be released is on the order of 10^{12} Joules (J).

5.2.2 Radioactivity and Decay

Certain isotopes are considered *stable* and will remain in their atomic configuration indefinitely, whereas other isotopes are *unstable*, meaning they will decay through various particle emissions until they reach a stable configuration. The property of atomic or nuclear instability is referred to as *radioactivity*, and radioactive isotopes are commonly referred to as *radionuclides* or *radioisotopes*.

The number of decays per second is referred to as the *activity* of a radioisotope and is measured with units of Becquerels (Bq) or Curies (Ci), where 1 Ci = 3.7×10^{10} Bq.

It is impossible to predict exactly when an individual radioisotope will decay, but the average rate of decay of a given atomic population can be known to a high degree of certainty. Therefore, a common term to describe decay behavior of isotopes is the *half-life*, (*T*) which is the time it takes for half an initial radioisotope population to decay.

The half-life (*T*) dictates the exponential decay behavior of a radioisotope population through a decay constant (λ) as shown in equation (5.1). This relationship can be used to predict the activity (*A*), thermal output (*W*), or number of atoms (*N*) of a given population of radioisotopes as a function of time (*t*) using equations (5.2–5.4).

$$\lambda = \frac{T}{ln(2)} \tag{5.1}$$

$$A = A_o e^{-\lambda * t} \tag{5.2}$$

$$W = W_o e^{-\lambda * t} \tag{5.3}$$

$$N = N_o e^{-\lambda * t} \tag{5.4}$$

A generic illustration of radioisotope decay is presented in Figure 5.2, where the *y*-axis represents the fraction of an initial activity remaining as a function time. To further highlight the context of half-lives in the decay process, the first (*1T*), second (*2T*), and third (*3T*) half-lives are marked. While the decay behavior illustrated in Figure 5.2 represents activity (*A*), this behavior can also apply to an isotopic concentration's thermal output and number of atoms.

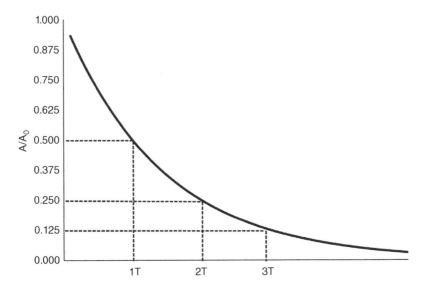

Figure 5.2 Exponential decay behavior of radioisotopes related to first three half-lives. Graph adapted from Turner's, *Atoms, Radiation, and Radiation Protection*.

Although half-lives of all the known isotopes can range anywhere from 1×10^{-24} seconds to 1×10^{24} years, it is typical that half-lives of radioisotopes relevant to RTGs are on the order of hundreds of days to hundreds of years.

5.2.3 Emission of Radiation

Discussion of isotopic activity in this chapter has been relegated to the generic notion that an atom simply *decays*. However, it is important to make the distinction that isotopes can decay by many different mechanisms, and these mechanisms fundamentally define their primary particle emission-types, potential secondary particle reactions, heat generation, and resulting decay products. While many of these phenomena include compelling physics topics relevant on a nuclear-scale, further discussion of radiation emission is limited here to interactions that influence radioisotope fuel evolution, heat generation, and radiological concerns that are relevant to macro-scale RTG operation and design.

When considering typical RTG fuels, four primary decay modes dominate relevant radiation phenomena: alpha (α), beta (β), gamma (γ), and spontaneous fission (SF). Some of these decay modes change or *transmute* the original isotope into a new isotope (i.e., α and β), others fragment an original isotope into multiple smaller isotopes (SF), and other (γ) only transitions an atom from one excited state to another, leaving the particle inventory of the isotope's nucleus unchanged. High-level comparisons between these decay modes and resulting particle emissions are presented in Table 5.2.

Although the primary decay modes and emissions shown in Table 5.2 are spontaneous, meaning that they do not require any incident particles or energy to occur, the distinction between these and nuclear reactions is worth noting. Reactions are different from spontaneous decay in that they will not transpire without an outside stimulus: they *require* an incident particle to occur. Examples of two reaction types relevant to RTG fuels are presented in Table 5.3. All potential

Table 5.2 Natural decay modes relevant to RTG fuels.

$^A_Z X_N$ $\rightarrow ^{A-4}_{Z-2}X'_{N-2} + ^4_2\alpha_2$	$^A_Z X_N \rightarrow ^A_{Z-1}X'_{Z+1} + \bar{v}_e + \beta^-$ $^A_Z X_N \rightarrow ^A_{Z+1}X'_{Z-1} + v_e + \beta^+$	$^A_Z X_N \rightarrow ^A_Z X_N + \gamma$	$^{A+B+C}_{Y+Z}X_N \rightarrow$ $^A_Y X' + ^B_Z X' + C^1_0 n$
Alpha (α)	Beta[a] (β^- & β^+)	Gamma (γ)	Spontaneous Fission[b] (SF)

[a] Symbols $\bar{v}_e$ & $\bar{v}_e$ represent anti-neutrino and neutrino emissions that occur during $\beta^\pm$ emission, which do not contribute to local heat deposition or radiation damage but do account for this decay mode having an energy distribution as opposed to a discrete energy like α and γ emissions.
[b] Spontaneous fission typically results in multiple fission fragments (i.e., smaller elements) and the emission of 2 or 3 neutrons. Other prompt and delayed γ and n emissions are common during the fission process as well.

Table 5.3 Nuclear reactions relevant to RTG fuels.

$^4_2\alpha_2 + {}^A_Z X_N \rightarrow {}^{A+3}_{Z+2} X'_{N+2} + {}^1_0 n$	$^1_0 n + {}^{A+B+C}_{Y+Z} X_N \rightarrow {}^A_Y X' + {}^B_Z X' + C {}^1_0 n$
Alpha-neutron (α, n)	Fission[a] (F)

[a] Induced fission requires one incident neutron in a specific energy range to be absorbed by a target nucleus, temporarily creating an unstable nucleus, which typically fissions into multiple fission fragments (i.e., smaller elements) and the emission of 2 or 3 neutrons. Other prompt and delayed γ and n emissions are common during the fission process as well. Induced fission is not a primary concern for RTG fuels but is relevant in certain nuclear considerations and thus is included here.

decay modes, reactions, and particle emissions of radioisotopes are not presented in this work: the interested reader can find detailed discussions of these topics in multiple works in the open literature.

5.2.3.1 Alpha Decay

Alpha (α) decay is a spontaneous emission from a parent isotope of an ionized helium nucleus and is commonly observed in isotopes more massive than bismuth, with a few exceptions in some lighter elements. As the name implies, the nuclear strong force is much stronger than the other known forces – nuclear weak, gravitational, and Coulombic. However, the nuclear strong force is comparatively short-reaching in its sphere of influence, dropping precipitously after a mere ~1–2 femtometers (~1×10^{-15} m). [2] The alpha decay process is fundamentally the result of a nucleus with a large enough proton and neutron inventory (i.e., heavy elements) to create a nucleus volume—and thus diameter—in excess of the nuclear strong force's overpowering influence. Once this distance threshold is exceeded, the repulsive Coulombic force between the positively charged protons in the nucleus dominate and accelerate these peripheral protons away from the more massive nucleus in a preferentially bound configuration of two protons and two neutrons known as an *α-particle*. This process brings the contents of the parent nucleus back within a smaller volume, which provides less opportunity for these types of events, and thus results in a more stable nuclear configuration. Because α-emission is a two-body decay process involving only the parent nuclei and the α-particle, an α-particle emitted from any parent isotope will have a predictable and discrete kinetic energy ranging between ~4–8 MeV yet will always have a 2^+ charge. The kinetic energy of the emitted α-particle is specific to each parent isotope, but typically, ~5 MeV is the approximate α-energy observed from PuO_2 fuels – which is primarily from ^{238}Pu. For comparison to other particles, a 5 MeV α will stop – or *range* out – in ~3–4 cm of air.[2]

---

2 https://physics.nist.gov/PhysRefData/Star/Text/ASTAR.html

5.2.3.2 Beta Decay

Beta (β) decay is a generic term describing various spontaneous nuclear emissions that can result when an unstable nucleus converts either a proton to a neutron or a neutron to a proton, to reach a more stable configuration. These processes result in the nuclear emission of an electron (with a -1 charge) or its anti-particle, a positron (with a +1 charge), dependent on the species of the parent isotope and decay mode. Atoms with more protons than neutrons will decay via $β^+$, and atoms with more neutrons than protons decay via $β^-$. Both β-decay modes create matter from energy during this transition, as no electrons naturally reside in the nucleus of an atom. When a β-particle is created, it is instantly imparted a kinetic energy as the result of a three-body reaction between the parent nucleon, the β-particle, and a neutrino.[3] As such, the exact kinetic energy of an individual β emission can never be predicted (like a two-body α-emission), but instead, will be represented by a spectrum of energies. However, the end-point energies of an emitted β spectrum, which are the highest energies achievable, are specific to each isotope, and the spectrum will range from zero to that end-point energy, with a mean energy at approximately ⅓ the maximum. [3] The charge of a β particle can be positive ($β^+$) or negative ($β^-$), depending on the mechanism of decay, but subsequent interactions of both particle types are treated largely the same outside of anti-matter annihilation reactions [3], which are not discussed here. For context, the range of a 1 MeV β-particle is approximately 4 m in air.[4]

5.2.3.3 Photon Emission

A photon is a massless unit of electromagnetic energy that travels at ~300×10⁶ m·s⁻¹ in a vacuum. Based on insights from Einstein, DeBroglie, and others, photons are defined as quantized packets of energy described by characteristics such as wavelengths, energy, or frequency, through the relationships shown in equation (5.5).

$$E = \frac{hc}{\lambda} \ \& \ \frac{c}{\lambda} = f \tag{5.5}$$

Where:

$E = Energy \ \left(J\right)$

$h = Planks \ Constant \ \left(6.6261 \times 10^{-34} J \ s\right)$

$c = Speed \ of \ Light \ in \ a \ Vacuum \ \left(2.997925 \times 10^{8} \frac{m}{s}\right)$

$\lambda = Wavelength \ \left(m\right)$

$f = Frequency \ \left(Hz\right)$

---

3 The neutrino is a small, neutral, subatomic particle important for describing the kinematics of β-decay, but largely unimportant to the macroscopic concerns of RTGs. [2] Further discussion of β-decay will omit discussion of neutrino (and anti-neutrino) emissions.
4 https://physics.nist.gov/PhysRefData/Star/Text/ESTAR.html

Photons are commonly thought of as visible light, but the electromagnetic spectrum represents many everyday phenomena, including AM and FM radio waves, microwaves, and ultraviolet waves, all of which are simply different energetic states of the same thing—photons. However, photons in the context of radioisotopes are relegated to the shorter wavelength, higher energy, higher frequency region of the EM spectrum, specifically, x-rays and γ-rays.

Photon emissions do not necessarily indicate a transmutation of the parent atom as seen in α and β decay. Transitions of atomic electrons between discrete orbital positions will emit corresponding discrete photons. Outer electron orbitals require less energy to excite or de-excite, so they release less photon energy during these transitions, whereas inner orbitals, specifically K and L shells, are responsible for higher energy photon emissions, which are known as *x-rays*.

The nucleus of an atom has an energetically stratified structure like that of electron orbitals, in that there are discrete energy states that a nucleus may experience while remaining as the same element. A nucleus can transition between these different discrete nuclear states by gaining or losing energy, as evidenced by the emission of discrete γ-rays. Also, most decay mechanisms for atoms (i.e., α, β, and SF) leave the nucleus (or the entire atom) in an excited state, which may subsequently de-excite by emission of γ-rays or x-rays. Once a nucleus has de-excited (perhaps multiple times) to a stable configuration via γ-emission, the nucleus is referred to as reaching a *ground state* (as opposed to an *excited state*).

γ- and x-rays are highly penetrating particles because of their massless state, neutral charge, and relatively high energy. Photons with energies as low as ~10 eV can still induce ionization (i.e., ultraviolet light), but the typical energies of photons associated with macroscopic concerns of RTG fuels are on the order of keV– MeV. For perspective, a population of 1 MeV photons will typically attenuate by a factor of ~1,000 in approximately 1,000 m of air.

5.2.3.4 Neutron Emission
5.2.3.4.1 Spontaneous and Induced Fission
Neutron emission from radioisotopes is typically a secondary product of some primary decay or reaction. For example, SF occurs when an unstable nucleus spontaneously fractures into two or more lighter elements, or *fission products*, while simultaneously emitting neutrons and photons. Also, induced fission can occur from a free neutron being absorbed in certain nuclei and causing a process similar to SF. The prospect of this occurrence is complicated, but it relies primarily on the energy of the incident neutron and the target atom's probability to undergo fission, as described by a *cross section* which is measured in cm^2 or *barns* (1 barn $= 10^{-24}$ cm^2). The resulting neutron energy spectrum from both spontaneous and induced fission can generically be represented by a Watt fission spectrum, with mean energies typically around 2– 3 MeV. [5]

5.2.3.4.2 (α – n) Reactions

This particular nuclear reaction is exotic in the context of everyday human activities but is quite common in the context of RTGs. The reaction occurs when a target nucleus absorbs a free α-particle and subsequently emits a neutron. Many α-decaying isotopes have a high thermal output, so they are commonly considered attractive for radioisotope fuels. Subsequently, these high-activity α emissions ($\sim 10^{10-20}$ s^{-1}g^{-1}) also provide ample opportunity for successive α-n reactions to occur in the RTG fuel, even in scenarios having low reaction probabilities. α-n reactions are more probable in low-Z materials (e.g., Li, Be, C, or O)[5] and are commonly found within RTG oxide fuel forms that have low, yet impactful, concentrations of oxygen isotopes such as ^{17}O and ^{18}O. [6,7] For example, in PuO$_2$ fuels, neutrons released from α n reactions (in non-purified oxides) can account for >80% of all the neutron emissions coming from the fuel. [6] Great time and expense are spent to reduce α-n reactions in PuO$_2$ by reducing non-^{16}O isotopes, along with other low-Z impurities, to make emitted neutron fields more manageable for radiological considerations during processing, shipping, handling, assembling, and spacecraft integration. Neutron emissions from α-n reactions have a complicated spectrum resulting from dependence on the energy of the incident α-particle and the nuclear inventory of the target isotope.

5.2.3.5 Decay Chains

Certain methods of decay, such as α, β, and spontaneous fission, described in this section alter the original or *parent* atom to form a new *daughter* atom. If a given daughter is also unstable, then it will decay (according to its unique half-life and decay modes) to another atom, and so on, until a stable nuclear configuration is achieved. This series of decays is referred to as a *decay chain* and is an important concept when discussing RTG fuels.

Isotopically pure samples will have a readily predictable behavior over time, but more realistic isotopic compositions may include several initial radioisotopes, all of which have independent decay chains. This makes for a complex and convoluted radiation emission spectrum and bulk decay characteristics. These respective decay chains will continue until stable isotopes are reached. An example decay chain of ^{238}Pu is shown in Figure 5.3, in which a series of decay processes result in a stable configuration of ^{205}Tl and ^{206}Pb.

5.2.4 Interactions of Radiation with Matter

So far, the discussion of radiation has been limited to phenomena associated with emission from the parent atom alone. However, these released particles are free to interact with matter around them, and all particles do not interact with matter in

---

5 Common examples of α-n reactions are ^{9}Be(α,n)^{12}C and ^{18}O(α,n)^{21}Ne.

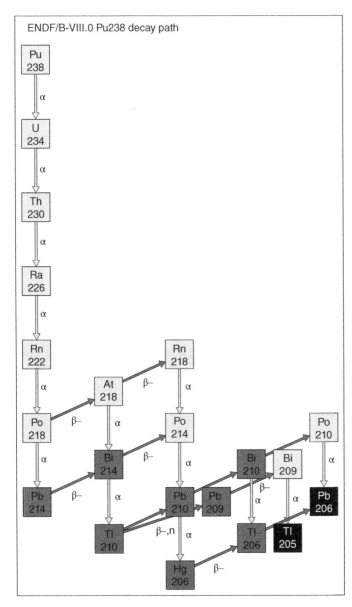

Figure 5.3 Decay chain of pure ^{238}Pu. Image permissions: OECD NEA, JANIS and the National Nuclear Data Center (NNDC) Evaluated Nuclear Data File (ENDF)/B-VIII.0.

the same way. Furthermore, the way that particles interact with matter is affected by the nature of the matter itself. For example, significantly different outcomes can result when two different particles interact with the same material or when two identical particles interact with two different materials. This section discusses some of the relevant phenomena associated with these interactions as they relate to RTGs.

When a free particle has enough energy to excite a target atom such that electrons are liberated, or *ionized*, then it is considered *ionizing radiation*. If there is not enough energy to ionize a target atom, then this is considered *non-ionizing radiation*. Although isotopes do generate thermal energy (non-ionizing EM radiation), the discussion of *radiation* in this section is referring to the emission of higher energy ionizing radiation on the order of keV–MeV.

5.2.4.1 Charged Particle Interactions with Matter

Charged particles such as α and β primarily deposit energy into matter via interactions with the Coulombic fields of atoms (i.e., via electrons and protons). One analytical model of the passage of charged particles through materials is described in the Bethe-Bloch equation and is commonly related to a concept known as *stopping power*. [3], which is a function of the incident particle's energy, mass, and charge, as well as the target medium's excitation energies and bulk electron density. In short, stopping power is a measure of how effective a specific material is at slowing down a specific charged particle. Figure 5.4 shows an illustration of how Coulombic forces cause a cloud of electromagnetic fields to which a charged particle is exposed when traversing matter.

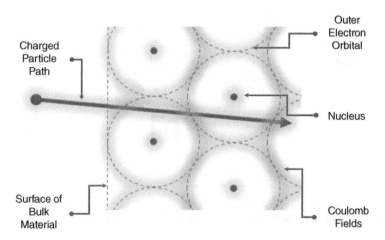

Figure 5.4 Illustration of the fields that charged particles experience while traversing matter. It should be noted that true charged particle paths through matter are not perfectly straight, and this example is only provided to highlight the large volume of Coulomb fields despite the small volume of actual matter.

The concept of *slowing down* a charged particle in matter is critical when understanding the fundamental mechanism of heat generation from radioisotope decay. The law of conservation of energy states that energy cannot be destroyed, but it can change form. As a charged particle traverses a sea of atoms, it slows down by losing its kinetic energy, which is directly converted to thermal energy deposition in the matter around it. This is relevant to RTG fuels because all α- and β-emitting isotopes emit particles that convert most of their initial kinetic energy back into heat in and around the vicinity of the very fuel where they were born.

As a simplified working example, consider an α-emitting fuel like PuO_2 and its primary isotope ^{238}Pu having a 100% probability of emitting (on average) a ~5.48 MeV α-particle during every decay. This is not a significant amount of energy on its own (5.48 MeV = ~8.8×10^{-13} J), but when one considers that value coupled with an alpha decay rate of ~10^{11} $s^{-1}g^{-1}PuO_2$ and the fact that a current RTG system holds ~4,800 g of PuO_2, it is easy to see how kW levels of thermal energy can be generated from a radioisotope. For example, a PuO_2 fuel with a weight fraction f_{Pu} of plutonium of 0.88 and an isotopic weight fraction f_{238} of ^{238}Pu of 0.85, the specific power P from alpha emission alone can be calculated as the product of the decay rate per unit mass $\lambda N/M$ and the alpha energy E_D as shown in equation (5.6).

$$
\begin{aligned}
P &= \frac{\lambda N}{M} f_{Pu} f_{238} E_\alpha \\
&= \left(\frac{\ln(2)}{2.77 \times 10^9 \ s} \right) \left(\frac{6.022 \times 10^{23} \ mol^{-1}}{238 g \ mol^{-1}} \right) (0.88)(0.85) \\
&\quad \left(\frac{5.48 MeV}{6.24 \times 10^{12} \ MeV \ J^{-1}} \right) \left(\frac{1 W}{J s^{-1}} \right) \\
&\cong 0.4 \ W \ g^{-1}
\end{aligned}
\tag{5.6}
$$

The concept of thermal energy deposition from radiation in matter is closely tied to a fundamental metric of radiation known as *absorbed dose*, which is simply a measure of energy deposition per unit mass, which can be reported in many units. The base SI unit of absorbed dose is the gray (Gy), and older units are referred to as the rad, where 1 J·kg^{-1} = 1 Gy = 100 rad.

5.2.4.2 Neutral Particle Interactions with Matter

Neutral particles (neutrons and photons) do not interact by the same mechanisms as charged particles, because they do not directly deposit energy in matter through Coulomb interactions. Instead, neutral particles deposit energy only by imparting some or all their *primary* energy into *secondary* charged particles liberated in the interaction. For example, *primary* neutrons interact with the atomic nucleus creating *secondary* charged particles while *primary* photons primarily interact with the atomic field resulting in the expulsion of one of the atomic electrons or the production of an electron-positron pair. The secondary charged particles liberated in these interactions go on to deposit their kinetic energy via the methods described above. The energy released from neutral particle interactions is called the *kinetic energy released per unit mass*, or kerma, and is comparable to the concept of absorbed dose if the attenuating medium is large. [3]

5.2.4.2.1 Neutrons

Neutrons primarily deposit energy through collisions with the nuclei of atoms, and kinematic relationships show that lower Z materials have a higher probability of neutron energy transfer per collision than higher Z materials. The term *moderation* is used to describe the slowing down of neutrons. Imagine a billiards table on which the cue ball is a neutron, and the eight ball is a hydrogen nucleus (a single proton). A keen player can make a shot that transfers all the cue ball's (neutron's) kinetic energy into the eight ball (hydrogen), thus stopping the cue ball in a single interaction. Conversely, if that same table is reset, but the eight ball (hydrogen) is now replaced with a bowling ball (perhaps neodymium), then one can imagine that after a collision, the bowling ball is not affected much, and the cue ball continues while retaining most of its initial kinetic energy. While not a perfect analogy, in general, materials like water (primarily consisting of hydrogen – a low-Z element) are better neutron moderators than, steel, tungsten, or lead (higher-Z materials) due to these kinematic considerations.

Other factors governing neutron interactions with matter require revisiting the concept of the *cross section*, which is briefly discussed above in reference to neutron-induced fission reactions. All nuclei have an energy-dependent probability of undergoing a given reaction. This probability is represented as a cross section, and these reaction cross sections can be considered individually, or they can be combined into a *total* cross section. The size and related complexity of the target nucleus determine the characteristics of the cross section. The probability of a nuclear reaction can vary dramatically over small energy spans in perturbations known as *resonance regions*. Figure 5.5 shows the total neutron cross section for ^{1}H (Figure 5.5a) and ^{238}Pu (Figure 5.5b), indicating the stark difference in the complexity of reaction cross sections. Note the complex resonance regions for the ^{238}Pu plot (Figure 5.5b) represented by the abrupt spikes midway through the energy spectrum. As cross sections fluctuate in magnitude, a given neutron population traveling through matter will have a higher or lower probability of undergoing some reaction, thus affecting the overall effectiveness of a given shield and the energy deposited.

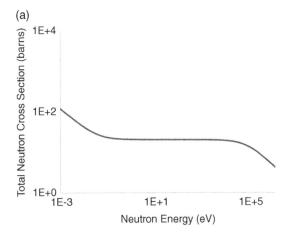

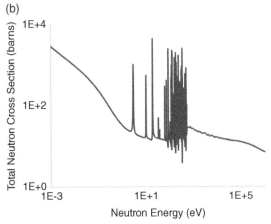

Figure 5.5 Total neutron cross sections for ^{1}H (a) and ^{238}Pu (b). Cross section data referenced from ENDF/B-VIII.0.

5.2.4.2.2 Photons

Photons are also neutral particles, but they primarily interact with electrons in matter through various scattering mechanisms. [3] While all photon interactions are not discussed here, the fundamental relationship of a material's average Z value and bulk density play significant roles in attenuation effectiveness of an incident photon field. As a rule of thumb, as average Z and bulk density of a given material increase, so does the effectiveness of the photon attenuation. Z is a driving factor, because the number of protons (Z) in a nucleus also indicates the number of electrons in a neutral atom, and electrons are the primary particle that photons interact with. Also, as density increases, the number of atoms per unit volume increases. The combination of Z and density indicate the overall electron density of a material and thus the effectiveness to shield and/or absorb incident photon energy.

5.2.4.3 Biological Interactions of Radiation with Matter

Previous discussions of radiation interactions with matter have been limited to fundamental atomic interactions in simple materials, not biological systems such as people. Typically, biological radiation exposures can be divided into two categories: deterministic (causing near-term radiation damage) and stochastic (resulting in long-term cancer development).

When considering deterministic radiation damage to humans, it is important to review specific definitions of ionizing radiation. All cells in a living organism are built on chemical bonds (ionic and covalent), and all chemical bonds are enabled exclusively by electron interactions. If a given atom is engaged in a chemical bond and the atom's outer electrons are ionized by an interacting particle, then the chemical bond might be broken. If this chemical bond is part of a living cell, then the breaking of that bond can decrease the cell's overall survivability. At low levels of ionizing radiation exposures, organisms may see no noticeable effects, whereas intense and acute radiation exposures can cause organisms to suffer significant cell death or even organism death. When considering large, acute radiation exposures in biological organisms, it is common to use an absorbed dose unit such as Gy or rad.

Whereas acute exposures can produce highly deterministic biological effects, it is much more complicated to predict stochastic effects like tumor formation. When analyzing stochastic radiation effects in complex organisms like humans, it is important to understand that new and complex genetic, epidemiological, and probabilistic variables must be considered. A significant exposure to radiation *now* can increase one's probability of cancer *later,* but not all extreme exposures lead to the same cancerous outcome: hence the stochastic aspect. The degree to which cancer development is a concern depends on multiple factors, including the type of radiation, the level of exposure, and what was irradiated (e.g., whole body, hands, head). The mechanism that links radiation exposures and potential cancer development originates from the fundamental interactions of radiation with a cell's deoxyribonucleic acid (DNA). If radiation ionizes or breaks a section of a given DNA chain, then the cell may be able to repair the damaged chain, but if the damage is irreparable, the cell may self-destruct to prevent future copies being made of the now flawed DNA. However, sometimes a radiation-damaged DNA chain can be miss-repaired, continuing to multiply as a mutation, thus becoming a cancerous colony of cells. [8] Manifestations of such phenomena may not present themselves until decades after a given exposure and predicting the likelihood of cancer development from a given exposure comes with substantial uncertainty. Because of the various factors that contribute to the increase or decrease in the probability of cancer forming from radiation exposure, a different dose metric is required. The concept of the *effective dose* (*E*) is used to measure radiological exposures for comparison to public and occupational limits. Determination of effective dose includes considerations like radiation type and tissue-weighting factors. [9] Effective dose is measured in units of sieverts (Sv) or rem, where 1 Sv = 100 rem, as described in equation (5.7), where D_j is the dose from radiation type j deposited in tissue i, w_{r_j} is the radiation weighting factor for radiation type j, and w_i is the tissue weighting factor for tissue i.

$$E = \Sigma_i w_{t_i} \Sigma_j w_{r_j} D_j \tag{5.7}$$

Radioisotope fuels used in RTGs are selected primarily based on their safety considerations, half-life, and thermal output. However, these nuclear characteristics come at the cost of also generating radiation emissions that present elevated, yet manageable, radiation environments in and around an RTG that must be considered.

β-emitting isotopes present a unique phenomenon during attenuation in which Coulombic forces exerted by atomic nuclei cause these passing charged particles to loose momentum. This abrupt momentum change will cause a β to emit photons as a mechanism to maintain conservation of energy [3] and the higher the average atomic number of the nucleus (i.e., the number of protons in a nucleus), the higher the photon yield from a passing β. This phenomenon is referred to as *bremsstrahlung* or *braking radiation*, and it presents a unique radiological concern for RTGs that use β-emitting isotopes. Although β particles themselves are relatively easy to shield, the production of bremsstrahlung can make certain RTG designs prohibitive because of the additional mass required to shield against the secondary photons.

α-emitting isotopes can easily be attenuated by mere distance (in air) or by adding minimal shielding. However, if α-emitting isotopes are consumed by a person or animal through inhalation or ingestion of particulates, then the radiological consequence can be significant. As such, great effort in design and testing is expended to ensure that α-emitting RTG fuels are in compounds and material encapsulations are resistant to atmospheric dispersion and solubility in water. [10,11] Furthermore, α-emitters are typically heavy atoms, which also presents opportunities for spontaneous fission or induced fission reactions, as well as the probability of inducing α-*n* reactions (see Section 5.2.3). These scenarios introduce free neutrons into the emitted radiation field and present an additional radiological concern that must also be mitigated.

Traditional spaceflight RTGs are mass constrained, making isotopic fuels that require significant shielding less attractive. As such, fuels typically considered for spaceflight RTGs have a comparatively low external radiation field when compared to many terrestrial nuclear materials. For example, a fully fueled multi-mission radioisotope thermoelectric generator (MMRTG) (~4.8 kg PuO_2) at beginning of life will present a radiological environment of ~10 mrem·h^{-1}at a distance of ~1.8 m (~6 ft). [12] For context, one hour of this level of exposure (at ~10 mrem·h^{-1}) is a comparable radiation dose to that received during a single commercial airline flight from New York to Tokyo (~7 mrem) [13] or during a single chest x-ray[6] (~10 mrem). Furthermore, the US Nuclear Regulatory Commission (NRC) guidance on radiological occupational exposures states that the annual limit for occupational workers is 5,000 mrem·y^{-1}. [14] While RTGs do present an elevated radiation environment when compared to natural background radiation, the intensities are quite manageable and exposures to personnel are safely limited by time, distance, and shielding to be well within typical nuclear working standards and practices.

5.3 Historic Radioisotope Fuels

There is no question that ^{238}Pu is the ideal radioisotope for heat generation for most space exploration activities. As an alpha emitter, nearly 100% of the energy released from ^{238}Pu during radioactive decay is converted into heat. Its half-life is short enough to produce good specific thermal power (W·g^{-1}), but long enough to keep the RTG functioning for decades. An RTG design does not require any additional radiation shielding to make it safe for humans to handle for short periods of time and finally, the PuO_2 is in a chemically stable, and insoluble form that is reasonably easy to produce and process.

In addition to these key chemical, physical, and thermal properties, PuO_2 heat sources typically have a very high fraction of the desirable heat-producing isotope – ^{238}Pu. Freshly made heat source plutonium can contain over 90% ^{238}Pu. [15,16][7] Many other heat-producing isotopes have much lower isotopic purity levels, such as ^{90}Sr, which is usually obtained at purity levels around 50%.[8] In addition, PuO_2 is ~88% plutonium by mass, so a majority of the chemically stable form is the element plutonium. Compare this to ^{90}Sr again, which typically uses a $SrTiO_3$ fuel form, where only 49% of the mass is strontium. These factors are important when considering a heat source isotope. Whereas pure ^{90}Sr may have a theoretical,

---

6 https://www.health.harvard.edu/cancer/radiation-risk-from-medical-imaging
7 The remaining fraction is mostly ^{239}Pu with small but notable fractions of ^{240}Pu, ^{241}Pu, and ^{242}Pu.
8 Higher isotopic purity levels cannot be obtained by chemical processing. Isotopic purification methods are generally considered impractical and require very expensive, dedicated facilities such as calutrons, gaseous diffusion plants, or gaseous centrifuge facilities.

specific thermal power of ~1 $W \cdot g^{-1}$, an actual ^{90}Sr heat source typically only has a specific thermal power of ~0.2 $W \cdot g^{-1}$.

Despite ^{238}Pu being a clear favorite on paper, practical considerations have often led to consideration of non-^{238}Pu heat sources. One example is the development of an ^{241}Am heat source by the European Space Agency (ESA). [17] Compared to 0.4 $W \cdot g^{-1}$ for PuO$_2$ with ^{238}Pu, pure ^{241}Am has lower specific thermal power (0.114 $W \cdot g^{-1}$ [18]) and comparatively elevated dose concerns because of its more intense gamma emissions. It would be prohibitively expensive to stand up a full-scale ^{238}Pu production program in Europe, but ^{241}Am is an isotopically pure natural decay product found in civil plutonium stockpiles, and the facilities to separate and handle ^{241}Am already exist in Europe. [19] The development of an americium-based heat source for ESA was therefore determined to be advantageous because of the significantly lower cost, and it is also easier and more practical to implement.

Likewise, practical issues with ^{238}Pu production led the United States to investigate several other isotopes during the early years of the RTG program. Cost and scale were significant concerns, which led the government to consider isotopes that can be harvested from the waste streams of spent nuclear fuel. Some of these isotopes were even deployed in terrestrial RTGs, where the additional mass for radiation shielding is not a significant concern.

Another hurdle for the use of ^{238}Pu was the availability of ^{237}Np, the precursor required to make ^{238}Pu. When the first RTG was built in 1954, ^{237}Np was an uncommon isotope that was very expensive and difficult to make in large quantities. ^{237}Np is a long-lived[9] waste product generated during the creation of weapons-grade plutonium, and the acceleration of the Cold War in the 1950s and 1960s produced an ample stockpile of ^{237}Np. With the precursor supply issue resolved, the Savannah River Site (SRS) in Aiken, South Carolina, produced the first batch of ^{238}Pu for the RTG program in 1960 and improved production scale and efficiency over the next several decades. [20]

All RTGs built for US space missions have used ^{238}Pu, and this trend is likely to continue for the foreseeable future. Over the course of its history, however, the RTG program has invested a considerable amount of time, money, and effort developing a wide range of isotopes as potential heat sources. Some of these isotopes were successful enough to be deployed in terrestrial or prototype systems, although others showed critical flaws very early during development. Therefore, it is worthwhile to briefly review the non-^{238}Pu isotopes that were considered by the RTG program.

Because of the elusive nature of this topic, this review should not be considered comprehensive. Very few RTG reports were published in the open literature, and

---

9 Half-life = 2.14 million years

most of the programs were classified at the time. Many of these reports have functionally been lost because (1) they are not available in electronic format, (2) they are not considered valuable enough to extract from the physical archives, and/ or (3) they were never declassified. The final outcome is a mix of publicly available technical reports. The survey presented here represents a best effort to collect the most mature, (e.g., programmatic final reports) publicly available information on other heat source candidates so that some level of the progress, successes, and failures can be recorded in a single place. Heat source candidates that appeared promising enough to advance to some level of development are surveyed here.

5.3.1 Polonium-210

^{210}Po was a natural choice for use in the early development of RTG systems and was made into a workable fuel form by mixing with a gadolinium alloy. At the time, ^{210}Po was relatively easy to obtain, required very little shielding for safe handling conditions, and the gadolinium alloy fuel form had an exceptionally high specific thermal power of 82 W·g^{-1}. [21] These properties made it very attractive as a heat source for development and demonstration of RTGs. Two famous examples are the first proof-of-concept RTGs ever made in 1954 and the 1959 Systems for Nuclear Auxiliary Power (SNAP)-3 that can be seen in the quintessential photograph of an RTG on President Eisenhower's desk (c.f. Chapter 1).

Unfortunately, the counterpart to high specific thermal power is a very short half-life of 138 days, which makes the isotope unattractive for many space missions. The SNAP program found ^{210}Po attractive for the SNAP-29 design, which was intended for use on satellites that required high power, low mass, and low environmental concerns from re-entry several years later. [22] SNAP-29 was never developed past the point of planning for ground tests. The Soviet space program also publicly reported[10] using the ^{210}Po Orion-1 RTG on satellites that have the National Aeronautics and Space Administration (NASA) designations of Cosmos 84 and 90, and ^{210}Po radioisotope heater units were used on the Lunokhod-I and Lunokhod-II missions to keep the rovers warm during the long lunar night. [23]

5.3.2 Cerium-144

^{144}Ce was the first radioisotope that was seriously considered for use in a mission generator, with several reports that predate the SNAP moniker. [24] ^{144}Ce is a highly abundant fission product in spent nuclear fuel, making it cheap and readily

---

10 The Soviet space program was notorious for not publicly releasing information. The missions discussed here were all publicly acknowledged by the Soviet Union, so it is possible that there were additional Soviet missions that used radioisotope heat sources.

available. In addition, the CeO_2-10% SiC-0.5% CaO fuel form has a high melting point, is insoluble in water, provided safe re-entry properties, and has a good specific thermal power of 1.96 $W \cdot g^{-1}$. [25] However, as a beta-emitter with a short half-life, the bremsstrahlung radiation required massive radiation shielding to enable direct human handling and interaction with the generator.

[144]Ce was considered as a heat source for the SNAP-1 mercury vapor Rankine cycle engine, the SNAP-1A PbTe RTG, and the SNAP-3 RTG. [25–27] All these designs included a 6-in. thick sphere of mercury for additional radiation shielding, with the notion that the mercury could be drained after installation and prior to launch. Ultimately, the short half-life of [144]Ce (285 days) was a major complication for these systems, as initial ground testing of a [144]Ce-based SNAP-3 showed that the RTG would lose 50% of its power after only 6 months. [27]

5.3.3   Strontium-90

[90]Sr is an attractive heat source because of its high yield in spent nuclear fuel, making it cheap and readily available. The mid-range half-life of 28.9 years would allow an RTG to operate for years or even decades before power degradation became a significant issue. Finally, the ~0.2 $W \cdot g^{-1}$ specific thermal power of the $SrTiO_3$ fuel form was high enough to make it a practical heat source. As a beta-emitter, the bremsstrahlung radiation did require additional shielding, but unlike [144]Ce, the half-life was long enough that the shielding requirements were more reasonable.

Because of these highly favorable characteristics, ^{90}Sr saw widespread used as a heat source for RTGs, particularly in terrestrial systems, where the additional mass for shielding was much easier to accommodate. Clearly, the most famous ^{90}Sr RTGs are the Russian "lighthouses" that were used as remote navigation beacons and weather stations. [28] It was estimated that over 1,000 of these RTGs were installed in remote regions of the Soviet Union between the 1960s and 1980s, and all of these generators outlived their 10-year design life. Reports estimate that a concerted international effort eventually removed over 80% of these generators from service [29], but a significant number of remotely installed radioisotope heat sources still remain.

The United States also developed and installed several terrestrial ^{90}Sr RTGs. [30] One generator, simply known as Sentry, powered a joint US–Canada remote arctic weather station. [31] The SNAP-7 series of RTGs (variants A through F) were designed as navigational aids, deep sea sonar stations, and automated weather stations. [24, 32–34] SNAP-21 was used to power a demonstration unit for an underwater oceanographic beacon. [35] SNAP-23 was an Atomic Energy Commission (AEC) R&D effort to improve efficiency, and although reports do not indicate that this system was ever deployed, it was notionally designed to use a ^{90}Sr heat source. [30,36] SNAP-17 is the only ^{90}Sr RTG that was being developed for space applications, and the selection of ^{90}Sr was justified as "… only this fuel is to be available in the necessary quantities." [37] Publicly available reports, however, do not discuss development of this system beyond a Phase I effort.

5.3.4 Curium-242

^{242}Cm was an attractive isotope for very short-term missions because its half-life of 163 days was similar to that of ^{210}Po but could be produced at ~3% of the cost. [38] The fuel form used to power demonstration units was a Cm_2O_3-AmO_2-Ir ceramic metal, or *cermet*, with a specific thermal power of ~6.7 W·g^{-1}, although much higher (or lower) specific thermal powers could be obtained by changing the amount of AmO_2 and/or Ir used. [39,40] Shielding needs were minimal, because the radiation profile has a slightly higher x-ray intensity but is otherwise similar to PuO_2 (per thermal watt).

^{242}Cm was used as the heat source for the SNAP-11 and SNAP-13 demonstration units. Both of these units were being developed for the NASA Surveyor Project, where the heat source was intended to provide at least 90 days of uninterrupted power and heat during the long lunar night. [41] The high specific power of ^{242}Cm was of particular interest during development of the SNAP-13 because the thermionic power conversion concept required heat source temperatures above 1,650°C. [42] ^{242}Cm fuel forms were never developed past the demonstration phase.

5.3.5 Curium-244

^{244}Cm was one of the most favorable alternatives to ^{238}Pu because of its relatively low cost to produce and the ability to make into a Cm_2O_3 fuel form (oxide fuels have relatively high melting points and are insoluble in water). Furthermore, its high specific thermal power of 2.00 W·g^{-1} was attractive for use in thermionic generators, which were estimated to be over 50% more efficient than any RTG that has ever been produced. [38] A report from 1970 indicates that the cost to produce a thermal watt of ^{244}Cm was the same as that of ^{238}Pu, and with some investments, production costs could be reduced by 80%. [42] Radiation shielding requirements are higher for ^{244}Cm because it emits a neutron flux that is almost 1,000× higher than ^{238}Pu [39], but this additional shielding was not considered a detriment because the shielding mass was more than offset by mass decreases caused by the higher thermal density. [28,42][11] ^{244}Cm was a target for heat source development

---

11 For an uncrewed mission. See Ref 26 for details on comparing ^{244}Cm vs. ^{238}Pu shielding for a mission including humans.

for years, but it never received the investments necessary to make it a practical alternative to ^{238}Pu.

Estimates indicated that if the Multi-Hundred Watt (MHW) RTG had been fueled with ^{244}Cm, the power-to-mass ratio could have been increased by 10–15%. While this modest increase sounds good, it is fortunate that ^{244}Cm was never developed for this purpose. The shorter half-life of 18.1 years would have made the Voyager missions defunct decades ago, and they would not have been able to explore the outer solar system or any of the space beyond our solar system. Today, ^{244}Cm may still be a useful isotope for specialized missions that require very high thermal power density such as an ocean worlds probe that must melt through kilometers of icy crust. [29,30,43,44] However, many of the missions being planned today that would use an RTG would find the 18.1-year half-life too short to be useful.

5.3.6   Cesium-137

^{137}Cs was briefly considered as a heat source material because it is a plentiful fission product found in spent nuclear fuel with a reasonable half-life of 30.2 years. In addition, the cost per thermal watt of ^{137}Cs was estimated to be only ⅓ the cost of ^{90}Sr. As a beta-emitter, the shielding requirements were also estimated to be similar to those for a ^{90}Sr heat source. Unfortunately, the chemical reactivity and solubility of Cs are very high. This resulted in the development of a vitrified[12] Cs_2CO_3 fuel form. The large quantities of glass needed to contain Cs reduced the specific thermal power down to ~0.07 W·g^{-1}, and solubility tests showed greater than 10 ppm of Cs leaching into the surrounding water. [31,45] Concerns over low thermal power density and potential dispersal into the environment prevented ^{137}Cs from being developed past this initial fuel form study.

5.3.7   Promethium-147

^{147}Pm was investigated as a heat source material because the beta-emissions produced by the isotope are low energy, so a majority of the bremsstrahlung radiation is already shielded by the materials used to contain the heat source. ^{147}Pm is typically obtained through separation and purification from spent nuclear fuel or through irradiation of enriched ^{146}Nd targets. Both pathways are quite expensive. Once the isotope has been obtained, Pm_2O_3 is chemically easy to produce, has low chemical reactivity, and is insoluble in water. A specific thermal power of 0.23 W·g^{-1} was measured from a Pm_2O_3 heat source obtained from

---

12 i.e., dissolved in SiO_2-glass

spent nuclear fuel. [32,46] The half-life of 2.62 years is long enough to be attractive for some missions, but too short to be practical for nearly all space applications.

The cost and difficulty of obtaining [147]Pm was the primary limiting factor in developing this isotope as a heat source. Despite the costs, some RTG concepts were developed enough to produce a detailed design that would make use of the low radiation [147]Pm$_2$O$_3$ heat source for a short-term space mission. [32,46] Another development program progressed far enough to actually deliver a 60 thermal watt heater unit to the customer to provide heat for their mission. [33,47] Today, the cost, lower power density, and short half-life would preclude [147]Pm from being considered for most space applications.

5.3.8 Thallium-204

During the very early days of heat source development, [204]Tl appeared to be an exceptional isotope for heat source applications. It has a 3.8-year half-life, calculations suggested a specific thermal power of 1.04 W·g^{-1}, radiation consisted of medium-to-low energy beta-emissions, and it appeared to have a straightforward pathway for production using natural thallium. Unfortunately, initial irradiation testing on natural thallium indicated that it would take 360 days in a reactor to produce a sample with a specific thermal power of 0.0044 W·g^{-1}. [48] These results made it clear that production of [204]Tl as a heat source was impractical.

5.4 Producing Modern PuO$_2$

From the early 1960s through the early 1990s, [238]Pu was produced at SRS using large production reactors, canyon processing, and waste storage and solidification facilities [49], yet today, only a limited number of research reactors and radiochemical processing facilities remain available. At Oak Ridge National Laboratory (ORNL) in Oak Ridge, Tennessee, initial operations have begun to produce [238]Pu by irradiating [237]Np in the ORNL High Flux Isotope Reactor (HFIR). Plans are to use the Idaho National Laboratory Advanced Test Reactor (ATR) to supplement production. The transmutation path to produce [238]Pu from [237]Np is shown in Figure 5.6.

The heat source (HS) PuO$_2$ product specification requires the plutonium to contain at least 82.5% [238]Pu. [50] This means the [237]Np irradiation flux and time must be limited to transmute <15% of the [237]Np into plutonium, because [239]Pu, a major isotopic impurity, is produced by neutron capture of [238]Pu with a cross section that is ~3 times greater than the cross section for production of [238]Pu from [237]Np, Figure 5.6. The irradiated targets are processed at the ORNL

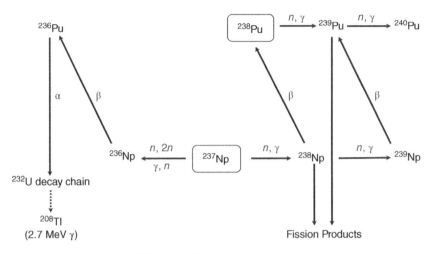

Figure 5.6 Production of ^{238}Pu from ^{237}Np.

Radiochemical Engineering Development Center (REDC) to separate and dispose of fission products and to partition and purify the ^{238}Pu product and unconverted ^{237}Np, which is recycled. The production process path is illustrated in Fig. 5.7. [51] In Steps 1 and 2, the makeup ^{237}Np feedstock, which was produced at the SRS, is dissolved, and then treated to remove ^{232}Th, which is an impurity in the feedstock, plus the ^{237}Np decay daughter ^{233}Pa has a 0.3 MeV gamma emission. Thus, the ^{233}Pa removal is necessary because process steps 3 and 4 (for conversion to NpO$_2$ and target fabrication, respectively), are carried out in unshielded or lightly shielded gloveboxes. Similarly, the recycled Np is treated for ^{233}Pa removal during chemical processing (Step 6). However, since the ^{233}Pa ($T \approx 26.9$ days) has a significantly shorter half-life than ^{237}Np ($T \approx 2 \times 10^Q$ years), the ^{233}Pa is readily replenished. Thus, each batch of purified ^{237}Np must be processed quickly to prevent excessive radiation exposure to operating personnel before significant ^{233}Pa buildup occurs.

The production process is a four-step rotating cycle (Steps 3 through 6 in Figure 5.7). One cycle is a production campaign. With current cermet targets, 63 targets are fabricated, irradiated, and processed in each campaign, and six to seven campaigns will be necessary each year to meet the production goal of 1.5 kg·y^{-1} of HS PuO$_2$. Scaling up the research and development facilities and operations at the REDC to meet the production goal requires annual irradiation and processing of several hundred targets containing a total of ~12 kg of ^{237}Np. The irradiation and processing facilities that are available normally irradiate and process significantly fewer targets at the gram scale and in a research and development mode. The scale-up and conversion of existing research facilities and operations to a

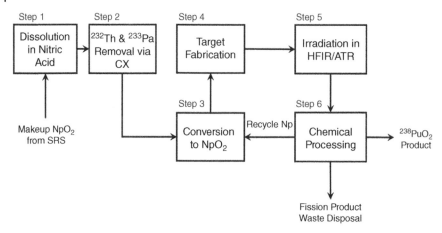

Figure 5.7 ^{238}Pu production process.

production mode pose significant challenges, but progress has been made. Increasing the rate of target fabrication is still one of the major challenges in the scale-up of the production process.

5.4.1 Cermet Target Design, Fabrication, and Irradiation

Aluminum-clad, aluminum-matrix metal oxide cermet targets are typically used for irradiation in research reactors such as HFIR and ATR. [52] This type of target design was chosen for ^{238}Pu production because it has been successfully used for many years for irradiation of plutonium, americium, and curium in ORNL's transuranium element production program. However, only 5–50 targets per year have been required for the transuranium element production rate needed, but for production of kilogram amounts of ^{238}Pu per year, coupled with the ^{237}Np limited to <15% Pu yield per irradiation cycle, multi-kilogram quantities of ^{237}Np must be irradiated each year to achieve this goal. This requires tens of thousands of pellets and several hundred targets to be fabricated, transported to and from the reactors, irradiated, and processed annually. [53]

In addition, the production rate is limited by the research reactor's operating time, the volume of available irradiation space in the reactor, and the ^{237}NpO$_2$ loading per target, which is limited by heat transfer and the 660°C melting point of aluminum. [15] Therefore, achieving the required ^{238}Pu production rate is challenging to say the least, and both HFIR and ATR will be required.

Regarding irradiation targets, about fifty pressed 0.25-inch diameter pellets are contained within the active length of the target, and each pellet volume contains 20% NpO$_2$, 70% aluminum, and 10% void space for fission gas expansion. The total weight of ^{237}Np in each target is 30.7 g. During the initial five campaigns, the

Table 5.4 Cermet target measuring performance.

Parameter	Parameter value or name	
Target Type	Cermet	Cermet
Number of HFIR cycles	2	3
Initial Np per target (g)	30.7	30.7
Conversion fraction to Pu	10.0%	12.6%
Isotopic purity of ^{238}Pu, by mass	88.5%	85.0%
Number targets in each ISVXF	7	7
Number of ISVXF positions	9	9
Sets of targets irradiated per year	3.50	2.33
Maximum number of targets irradiated per year	221	147
Number of targets for 1.5 kg PuO$_2$ per year	401	332
Irradiation rate in 9 ISVXF positions (g PuO$_2$/year)	823	643
Residual Np per year (kg/y)	9.6	7.3

targets have been irradiated for either two or three HFIR cycles. Irradiation performance is summarized in Table 5.4.

For the 63 target campaigns, the targets were irradiated in nine HFIR reflector positions, and each position contained an array of seven targets. The operating time for HFIR is seven cycles per year. Even though the conversion of ^{237}Np to ^{238}Pu is greater for the three-cycle targets, the isotopic purity of the ^{238}Pu produced is not as high, and more importantly, the number of targets that can be irradiated each year is greater for two-cycle irradiation (3.5 campaigns per year vs. 2.33 campaigns per year). Therefore, the annual yield of PuO$_2$ is also greater. [15] Even so, the yield is not enough to meet the production goal in HFIR, so the production must be supplemented with ATR irradiations.

To date, more than 500 targets containing >25,000 pellets have been successfully fabricated and irradiated. About 230 targets have been chemically processed in five campaigns.

5.4.2 Improved Target Design

Two approaches to improve target performance are being evaluated. First, thermal conductivity and thermal expansion tests are in progress to determine whether ^{237}Np loading can be increased. Initial results indicate an increase to 30% (vol.) loading is feasible and could possibly increase the HS PuO$_2$ yield by 20%. [54]

The greatest potential improvement is to modify the target design to eliminate the aluminum and to encapsulate NpO$_2$ pellets in a metal cladding such as

Zircaloy, which has a much higher melting point like that of commercial UO_2 fuel rods. Initial Monte Carlo N-Particle (MCNP) and SCALE modeling calculations were promising and were followed by test irradiations of Zircaloy-4 clad NpO_2 pellets that were 0.33 in. in diameter in an axial centerline HFIR irradiation reflector position. [55,56] Post irradiation examination showed that no melting, cladding interface problems, oxide swelling, shrinking, or otherwise deleterious effects occurred, even for four-cycle irradiation. [57] The isotopic purity of the [238]Pu was higher than that of the cermet targets, and conservative projections indicated that production of PuO_2 could be increased by a factor of ~2, and the number of targets required for fabrication and irradiation could be reduced by a factor of ~5. [15] In addition, the chemical processing steps to dissolve and dispose of aluminum waste would be eliminated. The only negative effect was that the amount of recycled [237]Np would be increased by a factor of 1.5, which is still acceptable.

5.4.3 Post-Irradiation Chemical Processing

The overall chemical flowsheet that is used to process 63 redundant target campaigns of irradiated cermet-form targets containing ~170 g of [238]Pu is shown in Figure 5.8. After dissolution of the aluminum cladding and pellet matrix in a caustic nitrate solution and subsequent dissolution of the actinide and fission product oxides in nitric acid, the primary separation of the plutonium product from neptunium and fission products will be accomplished through multistage solvent extraction using the existing mixer-settler equipment currently installed in the REDC hot cells for previous programs. The primary separation will be followed by (1) separate purification of the plutonium product and neptunium recycle material, (2) conversion of the plutonium product to the oxide form for shipment, and (3) recycling of the remaining neptunium for future target fabrication, as illustrated in Figure 5.7.

The success of the solvent extraction step, as well as the purification steps, is dependent on simultaneous control of the chemical valences of neptunium and plutonium. Significant success has been demonstrated for both coextraction of Np(VI) and Pu(IV) and for selective plutonium extraction of Pu(IV) from Np(V). Moreover, successful solvent cleanup and recycling to minimize organic waste generation in the solvent extraction step have been demonstrated. Purification methods for residual neptunium in the plutonium product and plutonium in the neptunium product have been demonstrated, and methods for removal of aqueous-soluble, phosphorus-based solvent degradation products have been demonstrated. Finally, a method for removal of thorium, which was a discovered impurity in the makeup neptunium feedstock, has been incorporated into the flowsheet, Figure 5.7.

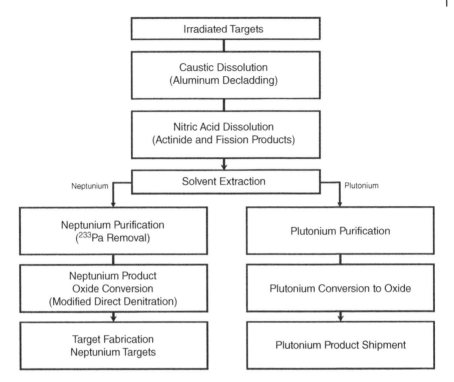

Figure 5.8 Chemical processing campaign flowsheet.

During the first five campaigns, target numbers 21–63 were processed to recover and purify approximately 50, 100, and 3×220 g of plutonium oxide product, respectively. The chemical processing operations have remained in the research and development mode to optimize the processes.

To meet or exceed the annual production goal of 1.5 kg of HS PuO_2, multiple two-month chemical processing campaigns will be required each year. Target dissolution will occur while components from the preceding campaign are separated. Scale-up to the operational level is expected to be accomplished by ~2024–2025.

5.4.4 Waste Management

The management of newly generated, high-activity liquid waste is an essential part of the processing campaigns. Efforts are being made to incorporate methods for liquid waste management into all radioisotope processing operations. Treatment of the accumulated radioactive liquid waste by extraction of traces of ^{238}Pu has allowed the waste to meet and exceed specifications for release. In addition, treatment of the alkaline aluminum liquid waste by adsorption of ^{238}Pu onto monosodium titanate has been successfully demonstrated and will be used in future campaigns to further reduce the remaining ^{238}Pu concentrations.

5.4.5 Conversion to Production Mode of Operation

Figure 5.9 illustrates the time challenges for typical full-scale production of ^{238}Pu, and it also illustrates the process to maintain the base radioisotope production program to produce other transuranium elements, primarily ^{252}Cf, which now requires a production campaign once every two years. [53]

In addition, all radioisotope production facilities require an annual mainte-nance period, as well as downtime for required technical safety compliance checks. Finally, it is necessary to include time for treatment of liquid wastes and disposal of solid wastes. In the coming years, the REDC facility operations sched-ule will be optimized, and the necessary personnel will be added to accomplish a sustainable production program for these isotopes. It will be necessary to conduct a series of processing campaigns for the ^{238}Pu program, each to occur in a time-frame of a few weeks. It will be necessary to provide additional shift operating personnel and supervision to enable multiple simultaneous operations. Use of the available tanks and evaporators must be carefully planned, and the overlap of the ^{252}Cf and ^{238}Pu campaigns must be eliminated. In addition, the annual outages and waste treatment campaigns must be planned well in advance and must be coordinated to minimize production process interruptions. Finally, the entire chemical processing of irradiated targets, ^{238}Pu product shipment, and recycled target fabrication must be carefully balanced to avoid exceeding the facility safety basis and material inventory limits.

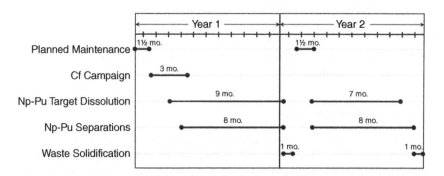

Figure 5.9 Potential full-scale production schedule for chemical processing at the REDC.

Efforts to develop an improved target design and fabrication rate, chemical separation processes, and waste management for the ^{238}Pu production project are continuing. Experience has shown that while successful production methods have been developed, significant challenges are still present that influence the ease at which each campaign can reach and maintain full production mode. While these challenges do exist, the current workflow of ^{238}Pu production is a proven method for successfully delivering high-assay and high-power density heat sources for the nation's space program. These methods will only be further optimized over the coming years to increase ^{238}Pu yield, throughput, and quality while decreasing waste, extraneous expense, and production time.

5.5 Fuel, cladding, and encapsulations for modern spaceflight RTGs

Although producing PuO_2 fuel for an RPS is a complex and niche field, the protective elements surrounding the fuel have similar levels of nuance and detail. Historic and current efforts to design, fabricate, test, and deploy protective claddings and heat shield components have been evolving since the advent of RTGs. The following sections present some of these engineering processes as well as discussing the methodologies for how these solutions protect the public from radiological concerns associated with RTGs during launch and spaceflight scenarios.

5.5.1 Evolution of Radioisotope Heat Source Protection

The first RTGs for space were developed under AEC's Systems for Nuclear Auxiliary Power (SNAP) program that originated at the Mound Laboratory. Engler et al. compiled a comprehensive historical account [58] of the evolution of radioisotope systems for space power. The SNAP-3B RTG was the first RTG to be deployed in space, providing supplemental power for navigational satellites. This RTG was soon followed by the SNAP-9A, which served as the sole source of power for satellites. [58] Both RTGs were powered by metallic ^{238}Pu fuel. Design and material selection for the SNAP-3B and SNAP-9A RTGs were based on meeting normal operational requirements. The safety design philosophy for these systems was known as a *burnup-dispersion* [59], where if a spacecraft inadvertently re-entered the Earth's atmosphere, the power system was designed to burnup during re-entry and disperse the radioisotope inventory into the upper atmosphere. This design approach was demonstrated when the third SNAP-9A–enabled satellite, Transit-5BN-3, failed to reach a stable orbit in

April 1964. Following this, the design philosophy shifted to what is known as *intact re-entry*, where the power system's radioisotope inventory does not burnup during re-entry and remains intact during decent and eventual impact with the Earth's surface.

The principal goal of the intact re-entry design approach was to protect the isotopic fuel within the heat source, although destruction of the thermoelectric converters and overall generator was expected during any re-entry scenario. This approach required integration of materials selection and design for normal operation with a broad range of accident scenarios that encompassed consideration of fuel form, fuel geometry, fuel spatial distribution, fuel cladding, chemical compatibility, thermal protection of the fuel during re-entry or launch events, ablation protection, and mechanical strength for incidents ranging from launch to re-entry to impact. Thermal management within the heat source is particularly challenging because the same materials and pathways that must transmit outbound heat to the hot side of the thermoelectric converters must also impede the potentially destructive inbound heat flux generated during re-entry.

It is instructive to review the SNAP-19 RTG heat source illustrated in Figure 5.10, which was the first to embody an intact re-entry design. [60] The fuel capsule was a singular stacked array of PuO_2 oxide and molybdenum cermet discs within a multi-layer primary containment consisting of a Mo-46Re inner liner, a Ta-10W liner, and a T111-Ta alloy strength member, all sealed within a Pt-20Rh cladding. Outside this fuel capsule, nested layers of pyrolytic graphite served as insulation. Enclosing all was an AXF-5Q graphite heat shield or *aeroshell*. The layers of refractory metal provided for high melting point containment. The Pt-20Rh alloy provided the final layer of containment and good oxidation resistance. [61] The pyrolytic graphite is a unique form of graphite that exhibits very high thermal conductivity within the plane of the cylindrical forms (axial and circumferential) and extremely low thermal conductivity through the thickness (radial). Additional spaces between these protective layers provided breaks in a direct conductive heat transfer path. Carbon materials such as fine-grained and pyrolytic graphite were prime candidates for ablation protection because carbon remains solid up to a higher temperature than any other element in the periodic table. Hence, graphite is well established in a multitude of very high temperature applications.

The SNAP-27 RTG was a contemporary of the SNAP-19, designed and deployed for Apollo Lunar Surface Experimental Packages. [62,63] The SNAP-27 heat source and RTGs were launched onboard six Apollo missions and made the journey as separate cargo within a graphite lunar module fuel cask on the Lunar Excursion Module. An astronaut inserted the heat source into the RTG housing

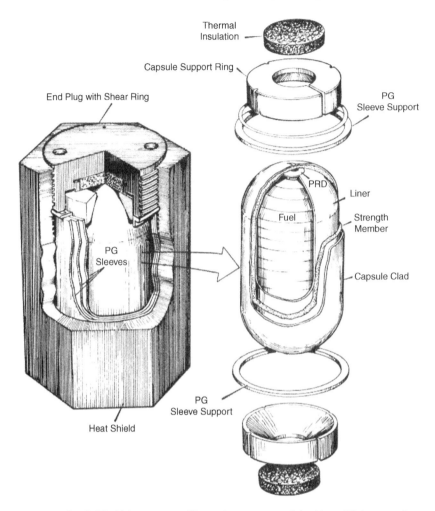

Figure 5.10 SNAP-19 heat source. Illustration courtesy of the Mound Science and Energy Museum Association.

on the moon's surface for five of those missions. The unique geometry of the SNAP-27 heat source is illustrated in Figure 5.11. [64]

The fuel was $^{238}PuO_2$ microspheres filling the annular volume between two co-axial superalloy cylinders. Both the inner cylindrical liner and the outer cylindrical clad were made of Haynes Alloy 25. Only one SNAP-27 heat source experienced a re-entry scenario. This occurred during the Apollo 13 mission due to equipment malfunctions preventing the mission from landing on the lunar

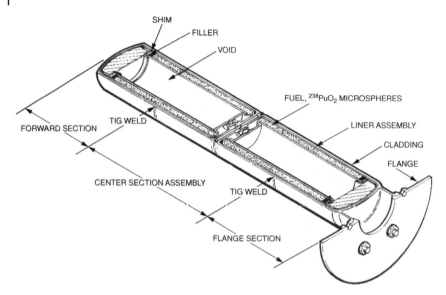

Figure 5.11 SNAP-27 heat source. Illustration courtesy of the Mound Science and Energy Museum Association.

surface, thus the power system never being deployed. Subsequently, the power system returned to Earth and endured an atmospheric re-entry without a detectable release of fuel. [65]

Next in the evolution of radioisotope power systems was the MHW-RTG, which served Earth orbit communication satellites and deep space missions on Voyager 1 and Voyager 2. This unique design is illustrated in Figure 5.12. [66]

The 24 fuel spheres for each heat source were fabricated by pressing $^{238}PuO_2$ powders into a spherical form. Each oxide fuel sphere was clad with 0.5 mm thick iridium metal as a primary containment. The fuel spheres were further enclosed within a carbon fiber sphere and an impact shell 13 mm thick. The carbon fibers were bound with a carbonized phenolic resin. The fuel spheres were introduced into the impact shell through a threaded cylindrical opening and were closed with a threaded partial-spherical segment. The 24 separate heat sources were arrayed within a cylindrical graphite cylinder with threaded graphite end closures that comprised the aeroshell. The MHW-RTG was the first high-power RTG to have an array of separate fuel components, each of which had stand-alone containment and protection from the spherical iridium cladding and impact shell. Under various re-entry scenarios, the carbon fiber sphere served a dual purpose as both an aeroshell and an impact shell. The thick wall of the carbon fiber sphere provided good ablation resistance, low radial thermal conductivity, and excellent impact crush resistance to accommodate the high terminal velocity expected from the spherical shape. [67]

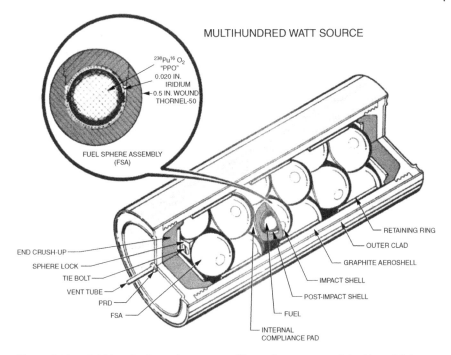

MULTIHUNDRED WATT SOURCE

Figure 5.12 Multi-hundred watt heat source. Illustration courtesy of the Mound Science and Energy Museum Association.

5.5.2 General Purpose Heat Source

The need for greater flexibility in RTG design and enhanced protection of the isotopic fuel in the event of a launch incident or from re-entry and Earth impact led to the development of a modular general purpose heat source (GPHS) design in the 1970s. [68] Each GPHS module contained four iridium-clad fuel pellets. Although each GPHS was assembled in arrays of 18 modules for the GPHS-RTG, or 8 modules for the MMRTG, each GPHS module provided for the requirements of normal operation and accident scenarios. Shock et al., summarized the creative evolution of the GPHS design as material selection, and detailed configuration options were quantitatively modeled and physically tested. [69] The central dilemma of the design process was to satisfy the competing requirements of normal operation and fuel protection. Under normal operation, the design and materials chosen must conduct and radiate heat to the surface of the aeroshell, where it is further transferred to the hot side of the thermoelectric generators (TEGs). The materials must also survive the high acceleration rates of various launch vehicles. An even greater challenge was to develop a design that could survive the broad range of accident scenarios. Launch explosions could expose the GPHS to fuel fires and impacts from an array of energetic vehicle fragments. Impact behavior of individual components and complete GPHS modules and module stacks were modeled and tested in a great variety of simulated impact scenarios. The design and materials had to provide high-impact strengths while also providing shock absorption for the ceramic fuel. The physical threats from a multitude of re-entry scenarios were also considered. Protection requirements varied if the re-entry was a launch abort from high or low altitude or re-entry from hypersonic gravity-assist flybys. The altitude, re-entry velocity, and angle of attack affect the ablation and thermal environment. The high-altitude hypersonic stage of re-entry generates high heat loads and high inbound heat flux, whereas the low-altitude subsonic stage of re-entry extracted heat to an extent that risked cooling the iridium clad below the ductile/brittle transition temperature. The first accepted design of the GPHS embodied the extraordinary ingenuity and effort, leading to an optimum compromise in achieving competing objectives. Some salient features of the GPHS design and materials of construction are discussed below.

5.5.3 Fine Weave Pierced Fabric (FWPF)

The materials of construction beyond the fueled clads are carbon because carbon remains solid to a higher temperature than any element in the periodic table. Carbon in the form of high-strength graphite as an ablative aeroshell and pyrolytic graphite as an insulator were employed to protect the isotopic fuel in legacy RTG heat sources. All structural components of the GPHS are made from fine weave pierced fabric (FWPF) carbon-carbon composite, including the aeroshell, the graphite impact shell (GIS), the floating membrane, the GIS cap, lock members, and lock screws. [70] FWPF is a high-density orthogonal weave carbon-carbon composite developed by Avco, now Textron Defense Systems, in the 1970s as an ablative re-entry material for the US Department of Defense (DoD). [71] The timely availability and extraordinary properties of FWPF led to its selection for the GPHS and aeroshell and components. The microstructure of FWPF is shown in Figure 5.13, adjacent to a legacy POCO graphite. [72] The three-dimensional character of the fiber reinforcement is evident. During fabrication, the matrix regions are filled by a carbon precursor and carbonized at high temperature through repetitive cycles, resulting in high density and sufficient bonding for composite behavior, in which the fibers carry most applied loads. The tensile strength of FWPF is approximately three times that of fine-grained graphite, and the thermal properties are similar, resulting in superior thermal shock resistance and markedly better ablation performance.

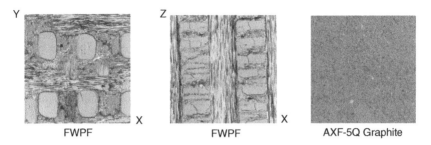

Figure 5.13 Aeroshell materials: fine weave pierced fabric and AXF-5Q graphite. US Department of Energy.

5.5.4 Carbon-Bonded Carbon Fiber (CBCF)

The design of the GPHS required a thermal insulator between the GIS and aeroshell to attenuate the inbound heat flux during hypersonic re-entry and high heat loss (cooling) during subsonic re-entry. Carbon-bonded carbon fiber (CBCF) was chosen to serve this role. CBCF insulation was first developed at the Y-12 National Security Complex and was further refined at ORNL for the GPHS application. [73] ORNL has produced CBCF insulators for the RPS program for over thirty years. A CBCF sleeve and two discs (end caps) envelop each GIS. The microstructure of CBCF shown in Figure 5.14 comprises chopped and carbonized rayon fibers bonded at the intersections by carbonized phenolic resin. The density of CBCF is

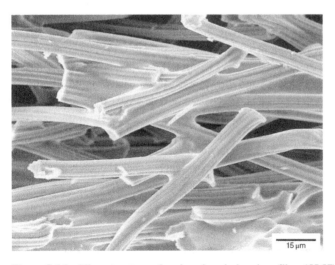

Figure 5.14 Microstructure of carbon-bonded carbon fiber (CBCF) insulation. US Department of Energy

0.20–0.25 g·cm^{-3} compared to 2.2 g·cm^{-3} for pyrolytic graphite. [69] CBCF was chosen over pyrolytic graphite for insulating within the GPHS to save mass and to help achieve the specific power goals for the GPHS-RTG. Each GIS with CBCF insulation is supported in the FWPF aeroshell at mating chamfers at the corners of the GIS cylinder. Consequently, the CBCF insulators receive essentially no external loading throughout the mission. GIS chamfers receive a thin coating of pyrolytic graphite to insulate this interface from conductive heat transfer.

5.5.5 Heat Transfer Considerations

A careful look at the geometry of the inside cylindrical surface of the GIS reveals several circumferential stand-offs to minimize the area of contact with the fuel cladding. Although the PuO_2 fuel pellet fits snugly within the iridium alloy clad with the internal weld shield, intimate contact is not ensured. The outer surfaces of the iridium alloy clads were grit blasted to enhance emissivity. After fueling and throughout launch, thermal management of the GPHS-RTGs (18 GPHS modules) was achieved using inert gas circulation and auxiliary heat rejection. After separation from the launch vehicle, the GPHS-RTG is vented to the vacuum of space. The principal mechanism of heat transfer within the GPHS-RTG components is radiative. In fact, the heat flux from the GPHS modules is radiated to the hot shoe of the SiGe unicouples. The cold side heat is radiated to space. Because the GPHS-RTG is operating at around 1,000+ °C, there is sufficient T, and ΔT to drive the required heat flux through the unicouples for power generation. The presence of low thermally conductive paths between the ablation surface of the aeroshell modules and the fueled clads provides sufficient thermal protection during re-entry scenarios.

The MMRTG adopted similar GPHS modules that were used for GPHS-RTGs. The MMRTG employs only eight GPHS modules in a thermoelectric generator designed for multi-mission applications, including Mars. The MMRTG operating temperature is only about 500 °C, and the average Martian atmosphere is approximately –65 °C, thus different thermoelectric materials were employed for these different operating conditions [74]. The heat source is sealed from the external environment by a Haynes 25 liner and operates under an internal environment of argon and helium gas. The high thermal conductivity and convective behavior of the gas serves as a primary heat transfer mechanism between the modules and the thermoelectric cavity.

5.5.6 Cladding

Starting in the early 1960s, RTG designs such as the SNAP-3B and -9B (Transits), SNAP-19B (Nimbus), SNAP-27 (Apollos 12, 14, 15, 16, and 17), and SNAP-19 (Viking and Pioneer) utilized tantalum, tantalum alloys, molybdenum-46%

rhenium, platinum-rhodium, and Haynes 25 in various combinations for fuel containment. [59–67] The MHW-RTGs used Ir-0.3 wt.% W alloy hydroformed into mated hemishells for primary fuel containment. [74,83–86] Note that the Ir-0.3 wt.% W alloy with the addition of 30 parts per million by weight (wppm) of Th, referred to as *DOP-4*, was used for the Voyager missions. [86,87] The Ir- 0.3 wt.% W materials (sheet and foil) were produced by ORNL in Oak Ridge, Tennessee, and all containment components were fabricated by the Mound Plant in Miamisburg, Ohio, hereinafter referred to as *Mound*.

The GPHS-RTGs used on Ulysses and Galileo [88], utilized the DOP-26 alloy containing Ir-0.3 wt.% W + 0.006 Th + 0.005 wt.% Al. The DOP-26 Ir alloy possesses a high melting point, fabricability, high-temperature strength/ductility, weldability, and chemical compatibility with the oxide fuel and the carbon materials that surround it. [89–91] ORNL produced the alloy by blending, pressing, and outgassing Ir and W powders. Seven compacts were melted multiple times with an electron beam to make buttons that were subsequently alloyed with Ir-Th and Ir-Al master alloys via arc melting. The alloy was drop cast to a $19 \times 19 \times 70$ mm long ingot and rolled to sheet to make blanks for subsequent two-stage clad vent set (CVS) cup-deep drawing operations at Mound. [87] This sheet production process is now referred to as the *old process* or the *previous process*. [92]

Cassini and subsequent missions (Pluto New Horizons, Mars Science Lab/ Curiosity, Mars 2020/Perseverance, and the future Dragonfly), regardless of RTG design (GPHS or multi-mission), utilize the same DOP-26 iridium alloy, but it was made using the current or *new* process at ORNL. Thus, blank production was scaled to a more efficient fabrication process [92–95] and was evaluated for fabricability [96], weldability [97], and high strain rate response. [98] The new process sheet fabrication differs from the old process in several ways. First, the arc-melted drop castings are 27 mm in diameter and 120 mm tall. After removal of any shrinkage cavity, typically nine of these cast segments are welded end-to-end with an electron beam to make an electrode that is nominally 90 mm long and subsequently vacuum arc remelted into an ingot that is 63.5 mm in diameter. The ingot ends are removed using a diamond grinding wheel and the cylindrical surface is ground for encapsulation in a machined-to-fit molybdenum can. The canned ingot is preheated to 1,425 °C for 3 hours then extruded to a rectangular sheet bar. The sheet bar is cut into sections. The molybdenum is dissolved using nitric acid and water plus sulfuric acid, and 1,200-gram sections are then rolled to sheet. Note that CVS production for the Cassini mission was conducted at the Y-12 National Security Complex, and subsequent CVS production was performed at ORNL. [99]

The vacuum hot-pressed PuO_2 fuel pellet is encapsulated in the iridium alloy CVS (see Figures 5.15 and 5.16) with an internal integral weld shield to avoid interaction between the cladding and fuel during the closure weld process. During production for the Cassini mission and subsequent missions, the CVS

Figure 5.15 Clad vent set: shield cup assembly at left with weld shield at top and vent cup assembly at right with decontamination cover at top. US Department of Energy.

Figure 5.16 Clad vent set: shield cup assembly in background with weld shield at bottom and vent cup assembly in foreground with frit vent assembly in bottom of cup. US Department of Energy.

vent notch area was increased to reduce the incidence of bulges and blowouts during the fueled clad closure weld operation. [100] The vent cup assembly, which is one of the finished CVS cups, has a vent hole that is nominally 0.45 mm in diameter centered in the bottom of the cup. The vent hole allows helium decay gas from the PuO_2 fuel to escape preventing pressure build-up inside the fueled clad. A frit vent assembly [101] is electron beam welded to the inside of the vent cup assembly over the vent hole allowing the helium to escape, but it prevents the escape of PuO_2 particles generated in a potential launch or re-entry event. A decontamination cover [102] is electron-beam welded to the outside of each vent cup assembly over the vent hole. This prevents post-closure-weld decontamination solutions from entering the vent hole. The decontamination cover is removed prior to assembly into a GPHS module. The outer surfaces of the CVS cups are grit blasted to enhance emissivity. The fueled clad temperature is maintained for thermoelectric performance during normal function and to ensure good ductility of the iridium alloy in potential accident scenarios such as those that might occur during launch or re-entry. [103]

Studies on the effects of impurities in iridium and alloy development efforts resulted in the DOP-40 alloy [104–108] containing 30 ± 10 wppm Ce and 40 ± 10 wppm Th. This alloy is comparable to DOP-26. It is available for use if radioactivity limits associated with the use of thorium are ever lowered.

Starting with production for the Cassini mission, fuel pellet production and encapsulation have been performed at Los Alamos National Laboratory (LANL). [109,110] For previous missions, this work was performed at the SRS, or even earlier at the Mound Laboratory. [111,112]

5.6 Summary

The benefits that spaceflight defense and science platforms provide is obvious to most and the technical complexities of these endeavors are unparalleled. However, developing an appreciation for the intricacies and evolution of the underlying RTG fuels and safety components that provide thermal power for many of these missions is the purpose of this chapter. The work of thousands of scientists, engineers, and technicians has culminated in the ability to provide a nuclear solution for remote, long-term, maintenance-free, reliable, and safe power generation in spaceflight applications. We hope the reader has enjoyed this information and gained a new appreciation for the nuclear complexities involved in the development of radioisotope fuels for RTGs.

References

1 Braibant, S., Giacomelli, G., and Spurio, M. (2012). The standard model of the microcosm. In: *Particles and Fundamental Interactions: An Introduction to Particle Physics* (ed. S. Braibant, G. Giacomelli and M. Spurio), 313–345. Dordrecht: Springer Netherlands https://doi.org/10.1007/978-94-007-2464-8_11.

2 Krane, K.S. and Halliday, D. (1987). *Introductory Nuclear Physics.* New York: Wiley.

3 Turner, J.E. (2007). *Atoms, Radiation, and Radiation Protection*, 3e Rev. and Enl. Weinheim: Wiley-VCH.

4 National Nuclear Data Center, information extracted from the NuDat database, https://www.nndc.bnl.gov/nudat/

5 Watt, B.E. (1952). Energy spectrum of neutrons from thermal fission of U-235. *Phys. Rev.* 87 (6): 1037–1041.

6 Matlack, G.M. and Metz, C.F. (1967). *Radiation Characteristics of Plutonium-238*. Los Alamos, NM: Los Alamos National Laboratory.

7 Taherzadeh, M. (1972). Neutron radiation characteristics of plutonium dioxide fuel. *NASA-CR-127045*.

8 Chu, E.H.Y. and Generoso, W.M. *Mutation, Cancer, and Malformation*, vol. 31. New York and London: Plenum Press https://link.springer.com/content/pdf/10.1007%2F978-1-4613-2399-0.pdf.

9 Eckerman, K., Harrison, J., Menzel, H.G., and Clement, C.H., and others (2013). ICRP publication 119: compendium of dose coefficients based on ICRP publication 60. *Ann. ICRP* 42 (4): e1–e130.

10 Reimus, M.A.H., Hinckley, J.E., and George, T.G. (1996). General-Purpose Heat Source: Research and Development Program. Radioisotope Thermoelectric Generator Impact Tests: RTG-1 and RTG-2. *Technical Report LA-13147*. Los Alamos National Laboratoryhttps://www.osti.gov/servlets/purl/279655 (accessed 19 August 2019).

11 Dahl, M., Supplemental Environmental Impact Statement for the Cassini Mission (1997). https://ntrs.nasa.gov/archive/nasa/casi.ntrs.nasa.gov/19990054126.pdf (Accessed 2 October 2017).

12 Smith, M.B., Peplow, D.E., Lefebvre, R.A., and Wieselquist, W. (2019). Radioisotope Power System Dose Estimation Tool (RPS-DET) User Manual. *ORNL/TM-2019/1249*, 1560442https://doi.org/10.2172/1560442.

13 Friedberg, W. and Copeland, K. (2003). What Aircrews should know about their Occupational Exposure to Ionizing Radiation. *DOT/FAA/AM-03/16*. Federal Aviation Administration.

14 10 CFR § 20.1201 Subpart C. Occupational Dose Limits, Code of Federal Regulations, 10 C.F.R. § 20.1201 (1991). https://www.nrc.gov/reading-rm/doc-collections/cft/part020/part020-1201.html (accessed 1 December 2022).

15 Collins, E.D., Morris, R.N., McDuffee, J.L., et al. (2021). Results, Implications, and Projections from Irradiation and Examination of Initial NpO_2 Test Targets for Improved ^{238}PU Production. Presented at the Nuclear and Emerging Technologies for Space (NETS), Virtual.

16 Matonic, J.H., Barklay, C.D., Rinehart, G.H., and Whiting, C.E. (2019). Chapter 31. use of plutonium in heat sources and radioisotope thermoelectric generators. In: *The Plutonium Handbook*, 2e, vol. 7. Clark, D.L., (ed.) American Nuclear Society.

17 Ambrosi, R.M., Williams, H., Samara-Ratna, P., et al. (2013). *Americium-241 Radioisotope Thermoelectric Generator Development for Space Applications*, 7. Recife: Brazil.

18 Oetting, F.L. and Gunn, S.R. (1967). A calorimetric determination of the specific power and half-life of americium-241. *J. Inorg. Nucl. Chem.* 29 (11): 2659–2664. https://doi.org/10.1016/0022-1902(67)80002-2.

19 Tinsley, T., Sarsfield, M., and Rice, T. (2011). Alternative radioisotopes for heat and power sources. *J. Br. Interplanet. Soc.* 64: 49–53.

20 Lindblom, K. (2021). *Plutonium-238 Production for Space Exploration - National Historic Chemical Landmark*. American Chemical Society https://www.acs.org/content/acs/en/education/whatischemistry/landmarks/plutonium-238-production.html.

21 Peterson, J., MacDonell, M., Haroun, L. et al. (2007). *Radiological and Chemical Fact Sheets to Support Health Risk Analysis for Contaminated Areas*. Argonne National Laboratory http://www.evs.anl.gov/pub/doc/ANL_ContaminantFactSheets_All_070418.pdf.

22 Teledyne, (1969). SNAP 29 Power Supply System. Ninth Quarterly Progress Report, *Report IND2062-12–8*. Isotopes Nuclear Systems Division. https://www.osti.gov/servlets/purl/4183615

23 Bennett, G.L. (1989). A Look at the Soviet Space Nuclear Power Program. In: *Proceedings of the 24th Intersociety Energy Conversion Engineering Conference*, vol. 2, 1187–1194. https://doi.org/10.1109/IECEC.1989.74620.

24 Raleigh, H.D. (1964). Systems for Nuclear Auxiliary Power (SNAP). A Literature Search. *Report TID-3561*. Oak Ridge National Laboratory, Oak Ridge, TN. https://doi.org/10.2172/4034092

25 Nuclear Safety Analysis Unit (1960). Preliminary Safety Analysis Low Power Cerium-144 Generator. *MND-P-2363*. Martin Company, Baltimore, MD.

26 P.J. Dick (1959). SNAP-1 Radioisotope-Fueled Turboelectric Power Conversion System Summary, January 1957 to June 1959. *MND-P-2350*. Martin Company, Baltimore, MD.

27 R.J. Wilson (1960). Conceptual Design of a SNAP-III Type Generator Fueled with Cerium-144. *MND-P-2369*. Martin Company, Baltimore, MD.

28 Sneve, M.K. (2006). *Remote Control*. IAEA Bulletin.

29 Diggs, C. (2015). *Foreign Funds Have Almost Entirely Rid Russia of Orphaned Radioactive Power Generators*. Bellona.org https://bellona.org/news/nuclear-issues/radioactive-waste-and-spent-nuclear-fuel/2015-11-foreign-funds-have-almost-entirely-rid-russia-of-orphaned-radioactive-power-generators.

30 Shor, R., Lafferty, R., and Baker, P. (1971). Strontium-90 Heat Sources. *Technical Report ORNL-IIC-36*. Oak Ridge National Laboratory, Oak Ridge, TN, https://technicalreports.ornl.gov/1971/3445605716035.pdf (accessed 6 May 2021).

31 Corliss, W.R. and Mead, R.L. (1963). *Power from Radioisotopes* (Rev.). US Atomic Energy Comission.

32 Anonymous (1961). Instruction Manual SNAP-7C Electric Generation System. *MND-P-2640*. Martin Company, Baltimore, MD. https://www.osti.gov/biblio/4673678

33 Young, C.N. (1963). SNAP-7D Strontium-90 Fueled Thermoelectric Generator Power Source Thirty-Watt US Navy Floating Weather Station. MND-P-2835. Martin Company, Baltimore, MD.

34 Palfrey, J.G., Ramey, J.T., Tape, G.F., and Seaborg, G.T. (1966). *Annual Report to Congress of the Atomic Energy Commission for 1965*. Washington, DC: US Atomic Energy Commission https://doi.org/10.2172/1364370.

35 Mandelberg, M. (1971). An oceanographic acoustic beacon and data telemetry system powered by a SNAP-21 radioisotope thermoelectric generator. *IEEE 1971 Conference on Engineering in the Ocean Environment*, San Diego, CA, 220–223. https://doi.org/10.1109/OCEANS.1971.1161004.

36 Weinig, J.F. (1970). SNAP-23A Program Thermoelectric Converter Development Program Final Report. *MMM-4187-0001*. 3M Company, Saint Paul, MN.

37 Streb, A.J. (1964). SNAP-17A System Phase I Final Summary Report, Part I. *MND-3307-33-1*. Martin Company, Baltimore, MD.

38 Weddell, J.B. and Bloom, J. (1960). *100-Watt Curium-242 Fueled Thermoelectric Generator – Conceptual Design*. Baltimore, MD: Martin Company.

39 Rimshaw, S.J. and Ketchen, E.E. (1969). Curium Data Sheets. *ORNL-4357*. Oak Ridge National Laboratory (ORNL), Oak Ridge, TN.

40 Jicha, J.J. and Haas, W.P. (1967). SNAP-13 Thermionic Development Program, Generator Fueling Report. *MND-3060-33*. Martin Company, Baltimore, MD.

41 Morse, J.G. (1963). Energy for remote areas. *Nature* 139 (3560): 1175–1180.

42 Carpenter, R.T. (1971). *Space Nuclear Power Systems*. Washington, D.C.: US Atomic Energy Commission and NASA.

43 Ambrosi, R.M., Kramer, D.P., Watkinson, E.J. et al. (2021). A concept study on advanced radioisotope solid solutions and mixed-oxide fuel forms for future space nuclear power systems. *Nucl. Technol.* 207 (6): 773–781.

44 Katalenich, J.A., Sholtis, J.A., Nesmith, B., and Fleurial, J.P. (2021). *Microsphere Plutonium-238 Oxide Fuel to Revolutionize New Radioisotope Power Systems and Heat Sources for Planetary Exploration*, 4. Pacific Northwest National Laboratory: Oak Ridge, TN.

45 Banks, H.O. (1961). The Cesium-137 Power Program, 3rd Quarterly Report. *RRC-0102*. Royal Research Corporation, Hayward, CA.

46 Nuclear Safety Analysis Unit (1959). Hazards Summary Report for a Two Watt Promethium-147 Fueled Thermoelectric Generator. *MCD-P-2049*, Martin Company, Baltimore, MD.

47 Davis, N.C. and Brite, D.W. (1969). Promethium-147 Radioisotope Application Program – AMSA Heat Source Final Report. *BNWL-994*. Pacific Northwest National Laboratory (PNNL), Richland, WA.

48 Blanke, B.C., Birden, J.H., Jordan, K.C., and Murphy, E.L. (1960). Nuclear Battery-Thermocouple Type Summary Report. *MLM-1127*. Mound Laboratory, Miamisburg, OH.

49 Wham, R.M., Felker, L.K., Collins E.D., et al. (2015). Reestablishing the Supply of Plutonium 238. Proceedings of the American Institute of Aeronautics and Astronautics (AIAA) Conference, Pasadena, CA.

50 Collins, E.D., and Wham, R.M. (2016). Development of Improved Targets, Separation Processes, and Waste Management for [238]Pu Production. Proceeding of the 2016 International Congress on Advances in Nuclear Power Plants (ICAPP), San Francisco, CA.

51 Collins, E.D., Morris, R.N., McDuffee, J.L., et al. (2021). Plutonium-238 production program results, implications, and projections from irradiation and examination of initial NpO_2 test targets for improved production. *Nucl. Technol. Am. Nucl. Soc.*, NETS Special Edition. 208: S18–S25

52 Bigelow, J.E., Corbett, B.L., King, L.J. et al. (1980). Production of Transplutonium Elements in the High Flux Reactor. Presented at the Symposium on Industrial-Scale Production-Seperation-Recovery of Transplutonium Elements, 2nd Chemical Congress. San Francisco, CA. (24–29 August 1980).

53 Collins, E.D. and Wham, R.M. (2018). Converting research and development facilities and operations into a [238]Pu production process. Presented at the Nuclear and Emerging Technologies for Space (NETS), Las Vegas, NV.

54 Collins, E.D. (2021). Private technical discussion.

55 Goorley, T., James, M., Booth, T., et al. (2016). Features of MCNP6. *Ann. Nucl. Energy* 87: 772–783. https://doi.org/10.1016/j.anucene.2015.02.020.

56 Rearden, B.T. and Jessee, M.A. (2016). *SCALE code system, ORNL/TM-2005/39, version 6.2.3* Oak Ridge Natl. Lab. Oak Ridge Tenn. Radiation Safety Information Computational Center as CCC-834, 2016.

57 Morris, R.N., Jordan, T., and Mulligan, P.L. (2021). Post Irradiation Examination of Candidate NpO_2 Targets for [238]Pu Production. *ORNL/TM-2020/1820*. Oak Ridge National Laboratory

58 Engler, R. (1987). Atomic Power in Space: A History. *DOE/NE/32117-H1*. US Department of Energy.

59 Secretary of The Atomic Energy Commission (AEC), (1963). Approval to use SNAP-9A on TRANSIT and AEC 1000/76 supplement to AEC 1000/75. AEC Memorandum to Director of Division of Reactor Development.

60 Teledyne Isotopes Energy Systems Division (1973). SNAP 19 Pioneer F & G Final Report. DOE/ET/13512-T1 report. DOE Final Tech. Rep. 156. https://www.osti.gov/servlets/purl/5352675.

61 Angelo, J.A. and Buden, D. (1985). *Space Nuclear Power*. Orbit Book Company Inc.

62 Pitrolo, A.A., Rock, B.J., and Remini, W. Snap-27 Program Review. In: *Intersociety Energy Conversion Engineering Conference Proceedings*, 153–170. American Institute of Aeronautics and Astronautics.

63 Remini, W.C. and Grayson, J.H. (1970). SNAP-27/ALSEP power subsystem used in the apollo program. In: *Intersociety Energy Conversion Engineering Conference Proceedings*, 13–15. American Institute of Aeronautics and Astronautics.

64 Prosser, D.L., (1969). SNAP-27 Radioisotopic Heat Source Summary Report. MLM-1698 report. AEC Research and Development Report, 12. https://www.osti. gov/servlets/purl/4758004.

65 *Response to Queries on SNAP-27 Reentry.* (1970). Director AAEC Division of Public Information.

66 *Mound Laboratory Data Sheet for Multi Hundred Watt.* Miamisburg, OH: Monsanto Research Corporation.

67 General Electric. (1978). Multi-Hundred Watt Radioisotope Thermoelectric Generator Program, Final Safety Analysis Report for the MJS-77 Mission. *77SDS4206.*

68 Snow, E.C. and Zocher, R.W. (1978). General Purpose Heat Source Development, Phase I Design Requirements. *LA-7385-SR.* Los Alamos National Laboratory, Los Alamos, NM.

69 Schock (1980). Design evolution and verification of the general purpose heat source. In: *Proceedings of the Intersociety Energy Conversion Engineering Conference,* American Institute of Aeronautics and Astronautics: 1032–1042.

70 Idaho National Laboratory. (2015). *Atomic Power in Space II – A History of Space Nuclear Power and Propulsion in the United States.*

71 Textron Inc. (2007). Textron Defense Systems Awarded $5.6M DOE Contract to Produce Fine Weave Pierced™ Fabric Carbon Carbon Billets For use by NASA. (23 February 2007).

72 Romanoski, G. (2004). GPHS Aeroshell Material Candidates – Salient Features.

73 Wei, G.C. and Robbins, J.M. (1985). Carbon bonded carbon fiber insulation for radioisotope space power systems. *Am. Ceram. Soc. Bull.* 64 (5): https://www.osti. gov/biblio/6091431.

74 Lee, Y. and Bairstow, B. (2015). Radioisotope Power Systems Reference Book for Mission Designers and Planners. *JPL-Publication 15–6.* Jet Propulsion Laboratory (JPL) https://ntrs.nasa.gov/search.jsp?R=20160001769 (accessed 18 September 2017).

75 Wyder, W.C., Powers, J. A., and Vallee, R.E. (1963). Destructive Evaluation of a SNAP Heat Source. *MLM-1180.*

76 Prosser, D.L. (1969). SNAP-27 Radioisotopic Heat Source Summary Report. *MLM-1698.*

77 Wendeln, D.E., Wyder, W.C., Tucker, P.A., and Newland, J.E.(1971). Source Capsule Fabrication Technology Program: September 1970-January 1971. *MLM-1796.*

78 Mound Laboratory. (1972). Mound Laboratory Isotopic Power Fuels Programs: April – September 1971. *MLM-1827.*

79 Wendeln, D.E., Coffey, D.L., Kreider, H.B.. (1972). Heat Source Capsule Fabrication Technology Program: Final Report January 1971 – March 1972. *MLM-1935.*

80 Selle, J.E. (1974). Advanced Heat Source Concepts. *MLM-2134.*

81 Skrabek, E.A. and McGrew, J.W. (1987). Pioneer 10 and 11 RTG performance update. *CONF-870102*, Albequerque, NM 1987, 201–204.

82 Bennett, G.L. (2006). Space nuclear power: opening the final frontier. 4th International Energy conversion Engineering Conference and Exhibit (IECEC) 2006–4191, American Institute of Aeronautics and Astronautics: San Diego, CA (June 2006), 17.

83 Braski, D.N. and Schaffhauser, A.C. (1975). Production of Ir-0.3% W Disks and Foil. *ORNL-TM-4865*. Oak Ridge National Laboratory (ORNL).

84 Wyder, W.C. (1976). Warm Hydroforming of Iridium + 0.3 wt % Tungsten Hemishells. *MLM-2203*

85 Liu, C.T. and Inouye, H. (1976). Study of Iridium and Iridium-Tungsten Alloys for Space Radioisotopic Heat Sources. *ORNL-5240*.

86 Liu, C.T. and Inouye, H. (1977). Development and Characterization of an Improved Ir-0.3% W Alloy for Space Radioisotopic Heat Sources. *ORNL-5290*.

87 Johnson, E.W. (1984). The Forming of Metal Components for Radioisotope Heat Sources. *MLM-3250*.

88 Bennet, G.L., Hemler, R.J., Whitmore, C.W., et al. (2006). The mission of daring: the general-purpose heat source radioisotope thermoelectric generator. Presented at the 4th International Energy Convention Engineering Conference and Exhibit.

89 Liu, C.T., Inouye, H., and Schaffhauser, A.C. (1981). Effect of Thorium Additions on Metallurgical and Mechanical Properties of Ir-0.3 pct W Alloys. *Metall. Trans. A.* 13A (6): 993–1002.

90 Liu, C.T. and David, S.A. (1982). Weld metal grain structure and mechanical properties of a Th-Doped Ir-0.3 Pct W Alloy (DOP-26). *Metall. Trans. A.* 13A: 1043–1053.

91 David, S.A., Miller, R.G., and Feng, Z. (2017). Welding of unique and advanced alloys for space and high- temperature applications: welding and weldability of iridium and platinum alloys. *Sci. Technol. Weld. Join.* 244–256. https://doi.org/10.1080/13621718.2016.1222255.

92 Ohriner, E.K. (1993). Improvements in manufacture of iridium alloy materials. In: *AIP Conference Proceedings*, vol. 1093, 271. American Institute of Physics.

93 Ohriner, E.K. (2008). Purification of Iridium by electron Beam Melting. *J. Alloys Compd.* 461 (1–2): 633–640.

94 Ohriner, E.K. (2008). Processing of Iridium and Iridium Alloys. *Platin. Met. Rev.* 52 (3): 186–197.

95 Ohriner, E.K. and George, E.P. (2011). Processing of high purity iridium alloys and effects of impurity elements. Presented at the Tungsten and Hard Alloys VIII, San Francisco, CA.

96 Forrest, M.A., McDougal, J.R., and Saylor, R.W. (1986). General Purpose Heat Source (GPHS) Clad Vent Set (CVS) Formability Study. *MLM-3395*.

97 Kanne, W.R.J. (1988). Weldability of general purpose heat source iridium capsules. In: *Proceedings of the Fifth Symposium on Space Nuclear Power Systems*, 279.

98 George, T.G. (1998). High-Strain Rate High-Temperature Biaxial Testing of DOP-26 Iridium Alloy. *LA-11065*. Los Alamos National Laboratory, Los Alamos, NM.

99 Ulrich, G.B., Ohriner, E.K., Romanoski, G.R., et al. (2014). *Heat Source Component Production for Radioisotope Power Systems*. Institute of Nuclear Materials Management: Atlanta, GA.

100 Ulrich, G.B. (1996). The Effects of Vent-Notch Area on Bulging and Thinning During the Clad Vent Set Closure-Weld Operation. *Y/DV-1425*.

101 Ulrich, G.B.(1994). The Metallurgical Integrity of the Frit Vent Assembly Diffusion Bond. Oak Ridge *Y-12 Plant, Y/DV-1321*.

102 Ulrich, G.B. and Berry, H.W. (1995). Summary of Decontamination Cover Manufacturing Experience. Oak Ridge *Y-12 Plant, Y/DV-1368*.

103 Schock, A. (1980). Design evolution and verification of the general purpose heat source. *Proceedings of the 15th Intersociety Energy Conversion Engineering Conference*, Seattle, WA (August 1980), 1032–1042.

104 Gubbi, A.N., George, E.P., Ohriner, E.K., and Zee, R.H. (1997). Influence of cerium additions on high-temperature-impact ductility and fracture behavior of Iridium alloys. *Metall. Mater. Trans. A* 28A: 2049–2057.

105 George, E.P. and Liu, C.T. (2000). Micro and Macro Alloying of Ir Base Alloys. *Proceedings of the 2000 TMS Annual Meeting*. Nashville, TN (12–16 March 2000), 3–14.

106 Heatherly, L. and George, E.P. Grain-boundary segregation of impurities in iridium and effects on mechanical properties. *Acta Mater.* 49: 289–298.

107 George, E.P., McKamey, C.G., Ohriner, E.K., and Lee, E.H. (2001). Deformation and fracture of iridium: microalloying effects. *Mater. Sci. Eng.* A319-321: 466–470.

108 McKamey, C.G., George, E.P., Lee, E.H. et al. (2002). Grain Growth Behavior, Tensile Impact Ductility, and Weldability of Cerium-Doped Iridium Alloys. *ORNL/TM-2002/114*. Oak Ridge National Laboratory.

109 E.A. Franco-Ferreira, M.W. Moyer, Reimus, M.A. et al. (2000). Characterization of Cassini GPHS Fueled Clad Production Girth Welds. *TM-2000/84*. Oak Ridge National Laboratory, ORNL.

110 Wham, R.M., Ulrich, G.B., and Lopez-Barlow, J.C. (2018). Constant Rate Production: DOE Approach to Meeting NASA Needs for Radioisotope Power Systems for Nuclear-Enabled Launches. *Am. J. Aerosp. Eng.* 5 (2): 63–70.

111 Groh, H.J., Poe, W.L., and Porter, J.A. (2000). Development and performance of processes and equipment to recover Neptunium-237 and Plutonium-238. In: *50 Years of Excellence in Science and Engineering at the Savannah River Site*, 165–178. Department of Energy.

112 Thomas Rankin, D., Kanne, W.R., Louthan, M.R. et al. (2000). Production of Pu-238 oxide fuel for space exploration. In: *50 Years of Excellence in Science and Engineering at the Savannah River Site*. Department of Energy.

6

A Primer on the Underlying Physics in Thermoelectrics

Hsin Wang

Oak Ridge National Laboratory, Oak Ridge, Tennessee

6.1 Underlying Physics in Thermoelectric Materials

The discovery of the thermoelectric effect [1–3] was almost 200 years ago. Today, solid-state refrigeration and power generation equipment use thermoelectric materials and devices. The history of the development of radioisotope thermoelectric generators presented in previous chapters shines a light on one of the outstanding achievements in thermoelectric technology. In addition, theories on thermoelectricity and the underlying physics have been improved over the past 60–70 years and are well documented. [4–10] This chapter presents a primer of the fundamental physics to help readers understand the essentials of past and current theories on thermoelectric materials related to RTGs.

All solid materials have charge carriers (electrons or holes), which are free to move when subjected to an internal or external potential field. Similarly, when a material, for example, a small bar, is subjected to a temperature gradient, the average energy of the free carriers at the hotter end of the bar increases, thus creating a carrier concentration gradient. However, the diffusion of the free carriers is counterbalanced by the development of an electric field, where the potential difference is known as Seebeck voltage. For a metal, the Seebeck coefficient, or thermal power, S is small, a few micro-volts per K. Equation 6.1 defines S mathematically:

$$S = \frac{\Delta V}{\Delta T} \approx \frac{\pi^2 k_B^2 T}{q E_F} \tag{6.1}$$

where ΔV is the Seebeck voltage across a temperature gradient, ΔT, k_B is the Boltzmann constant, T is the temperature in Kelvin, q is the electric charge, and

The Technology of Discovery: Radioisotope Thermoelectric Generators and Thermoelectric Technologies for Space Exploration, First Edition. Edited by David Friedrich Woerner.
© 2023 John Wiley & Sons, Inc. Published 2023 by John Wiley & Sons, Inc.

E_F is the Fermi energy, the highest state of electrons at 0 K. For non-degenerate semiconductors, thermal power is two to three orders of magnitude greater than in metals,

$$S = -\frac{k_B}{q}\left(A + \frac{E_C - E_F}{k_B T}\right) \quad \text{n-type}\left(\text{majority carrier : electrons}\right) \tag{6.2}$$

$$S = -\frac{k_B}{q}\left(A + \frac{E_F - E_V}{k_B T}\right) \quad \text{p-type}\left(\text{majority carrier : holes}\right) \tag{6.3}$$

where E_C is the energy of the conduction band and E_V is the energy of the valence band, and A is a constant depending on the carrier scattering mechanism. [11] The first part of equation (6.1) is the definition of the Seebeck coefficient, and equations (6.2) and (6.3) show the underlying physical properties and parameters using concepts of free electron gas and energy band. These simplified expressions for the Seebeck coefficient in semiconductors and their linkage to fundamental transport parameters will be discussed later.

The Seebeck effect is common to all solid materials. The most effective thermoelectric materials are semiconductors with carrier concentrations between 10^{19} to 10^{21} per cm^3. [12] In a circuit with a thermoelectric device, usually a couple with n and p legs, and an applied external load, current will flow when a temperature gradient is applied. Thus, the electrical conductivity, σ, of the element affects the level of power generation. The term, σS^2, is defined as the power factor (PF). To maintain steady power output in the circuit, the temperature gradient in the thermoelectric element must be stable. Thus, materials with a low thermal conductivity produce electricity more efficiently. There are many reviews in the literature [2-14] that stress the importance of the dimensionless figure of merit, ZT:

$$ZT = \sigma T S^2 / \kappa \tag{6.4}$$

where κ is thermal conductivity, and Z has the unit of 1/T, and T is the temperature in Kelvin. Equation (6.4) shows that to achieve maximum thermoelectric performance or higher ZT (dimensionless), the electronic transport properties or the PF must be maximized, and thermal conductivity, κ, must be minimized. This relationship underlines the importance of the transport of electrons and heat conduction in thermoelectric materials. In general, research to develop the "best" thermoelectric material has been ongoing for many decades, which has yielded incremental performance improvements. The standard for good thermoelectric materials is an integrated ZT close to 1.0 over the applicable temperature range. However, the selectable thermoelectric materials used by NASA and the DOE for

RTGs that power interstellar and planetary science missions have remained virtually unchanged in the last 50–60 years.

This chapter discusses the underlying physics that made thermoelectric materials such as PbTe/TAGS and SiGe the material of choice for RTGs and why no new materials have supplanted these flight-proven materials. But, first, it is important to discuss several solid-state physics concepts. It is critical to understand the theoretical analysis and experimental measurements of electron and thermal transport properties in semiconductors and thermoelectric materials.

6.1.1 Reciprocal Lattice and Brillouin Zone

Unlike metals in which the free electrons can be treated as a gas, semiconductors have electrons and holes in periodic potentials that follow more complex rules in transport theories. The discovery and advances in x-ray diffraction expanded our understanding of crystal structures and the periodicity of lattices in solids. Ions occupying periodic positions result in electrons seeing periodic potentials.

Bragg's law describes the x-ray diffraction of crystals, and a Fourier series conveniently presents the mathematical expression of periodic lattices in the momentum space. The diffraction pattern of a crystal projects a map of the reciprocal lattice in real space. Thus, the reciprocal lattice has a fundamental role in most analytic studies of periodic structures. The Fourier transform is the mathematical link between a crystal/direct lattice in real space and reciprocal lattice space. Bragg planes are defined as planes in the reciprocal space that bisect the lattice vectors at right angles.

In reciprocal space, the smallest repeating unit is called a primitive or Wigner-Seitz cell. The first Brillouin zone is a set of points that start from the origin without crossing any Bragg planes. For any crystal structure, the analysis of electron transport and lattice vibration in the first Brillouin zone represents the bulk material properties because of the symmetry of the lattice. This is an important concept for thermoelectrics because the electronic band structure and phonon density of state calculations only need to be carried out in the first Brillouin zone. Figure 6.1 shows the first Brillouin zones of three common crystal structures: simple cubic, face-center cubic, and body-centered cubic. The energy dependences in crystalline directions are identified using Greek letters, and the intercepts with the zone surface are labeled in Latin letters. For example, there are six high symmetry points (Γ, X, L, W, U, K) and six high symmetry lines (Δ, Λ, Σ, S, Z, Q) in the FCC first Brillouin zone.

6.1.2 Electronic Band Structure

While the classic free electron gas model can explain the transport behavior of metals, analyzing the electron waves using periodic potential in a crystal lattice

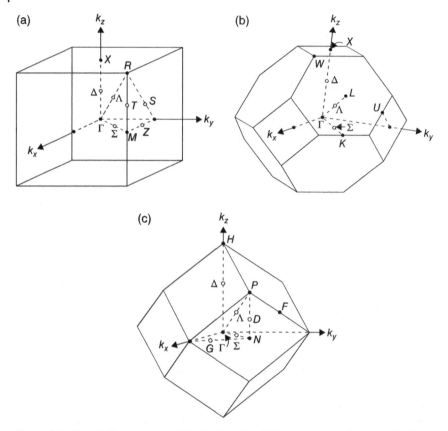

Figure 6.1 First Brillouin zones of (a) simple cubic, (b) face-center cubic, and (c) body-center cubic (with high symmetry points and lines).

becomes necessary to explain the transport properties of semiconductors and insulators. Figure 6.2a illustrates a 1-D lattice using the free electron model assuming an empty lattice. The Schrödinger equation (6.5) allows for the mathematical description of a free particle with mass m and zero potential:

$$-\frac{\hbar^2}{2m}\frac{d^2}{dx^2}\psi_k^{(0)}(x) = E\psi_k^{(0)}(x)$$

(6.5)

$$E_k^{(0)} = \frac{\hbar^2}{2m}k^2$$

(6.6)

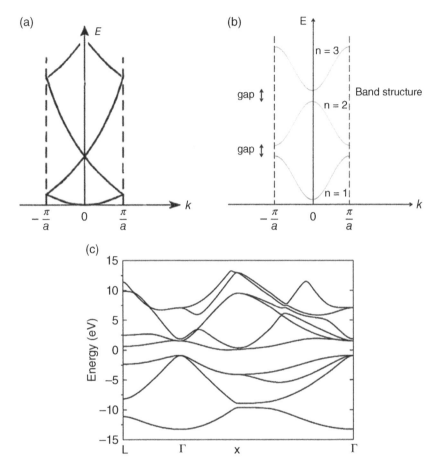

Figure 6.2 Energy vs. lattice directions for (a) Free-electron model and an empty lattice, no band gap; (b) 1D lattice with periodic weak potential, and a band gap at the zone boundaries; and (c) $Si_{50}Ge_{50}$ thermoelectric alloy lattice using linear muffin-tin orbital (LMTO) method. [15]

Equation (6.6) shows that E versus k follows a parabolic relationship with no energy gap. Using the nearly-free electron (NFE) model and a weak periodic potential (with lattice spacing a), we solve the Schrödinger equation again to determine the allowed energies. At the zone boundaries where $k = \pm n\varepsilon/a$ in Figure 6.2a, there are two possible energies; one is lower, and the other is higher than the free electron energy, E. Figure 6.2b shows the energy gaps of $2E$ where no electrons can exist. Using the same method in a 3-dimensional lattice for a thermoelectric material, e.g., $Si_{50}Ge_{50}$ alloy [15], Figure 6.2c shows the calculated energy band structure based on density function theory (DFT). Si and Ge have diamond cubic crystal structures, and SiGe is a solid solution of the two lattices. Γ (0,0,0), which is the origin in k space, L (½, ½, ½) and X (1,0,0) are high symmetry points, and lines connecting these points have minimum energy near Fermi level (Energy = 0 eV). A free simulation code developed by Madsen and Singh [16], BoltzTrap, is often used to calculate thermoelectrics' semi-classic transport coefficients and band structures.

Another important concept for band theory is the density of states (DOS), the number of orbital states, n, with energy between E and $E+dE$. If we examine the shape of the energy bands in Figures 6.2b and 6.2c, the shape of the $E(k)$ vs. the reciprocal lattice vector k is approximately parabolic (among lines of the three directions for $Si_{50}Ge_{50}$). They are upward at the bottom of the conduction band and downward at the top of the valence band. The effective mass, m^*, is the reciprocal of the curvature of the E vs. k plot and is used for electrons near the band edge and the density of states $N(E)$:

$$N(E) = \frac{V}{4\pi^2} \left(\frac{2m_b^*}{h^2} \right)^{3/2} E^{1/2} \tag{6.7}$$

$$N(E) = \frac{V}{4\pi^2} \left(\frac{2m_t^*}{h^2} \right)^{3/2} (E_{max} - E)^{1/2} \tag{6.8}$$

where V is the volume, h is Plank's constant, and m_b^* is the effective mass of an electron at the bottom of a conduction band, and m_t^* is the effective mass of the electron at the top of a valence band. The effective mass is also related to the electron group velocity, which affects the electrical conductivity. Figure 6.3 shows the calculated band structure by Chen [17] for several thermoelectric materials.

6.1.3 Lattice Vibration and Phonons

At any temperature above zero Kelvin, atoms in a solid vibrate about their equilibrium positions. When atoms are displaced from their equilibrium positions, forces acting on the atoms determine the elastic properties. Elastic waves that travel within a solid material are quasi-particles in quantum mechanics and are often called phonons. Phonons are quantized vibration modes of atoms bound into a material analogous to quantized sound waves, similar to photons as quantized light waves. Lattice vibrations generally occur at fixed frequencies, depending on the atoms and the crystal structure, ranging from GHz to several THz. Lower

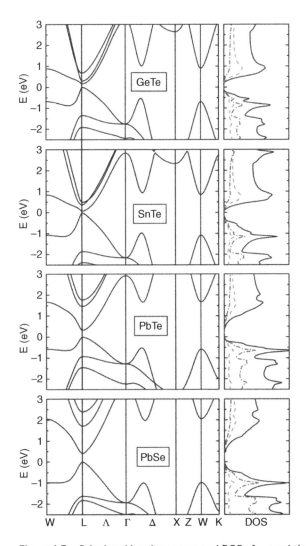

Figure 6.3 Calculated band structure and DOS of several thermoelectric materials. [17]

frequency modes (10-11 GHz) are dubbed acoustic phonons, representing atoms moving in phase. Higher-frequency modes (10-11 THz) are optical phonons, representing atoms moving out of phase. The term "optical" was applied to phonons because of the initial analytical technique used to observe higher-frequency phonons and because the electromagnetic spectrum associated with optical light also excites these phonons. Phonon dispersion relations are used to predict and calculate the thermal conductivity of solids. [18] A typical plot of Si phonon dispersion

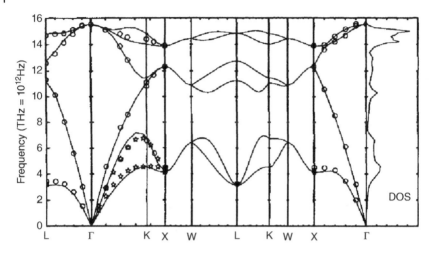

Figure 6.4 Phonon dispersion plot within crystalline Si. [19]

curves and density of state (DOS) is shown in Figure 6.4. [19] In thermoelectric power generation, low thermal conductivity is critical to maintaining a temperature gradient. Understanding phonon dispersion helps to understand the roles of crystal structure, defects, and dopants in optimizing thermal conductivity. Like the electron density of states, the phonon DOS can be calculated as a function of frequency by integrating the phonon dispersion curve. Thus, phonons travel at a group velocity, v, and the phonon mean-free-path is defined as the wavelength.

Thermal conductivity is defined as a heat/energy flux moving through a material under a temperature gradient:

$$Q = -\kappa \frac{dT}{dx} \tag{6.9}$$

and lattice thermal conductivity, κ_l is,

$$\kappa_l = \frac{1}{3}Cvl \tag{6.10}$$

where C is the heat capacity, v is group velocity (the slope of the dispersion curve), and l is the phonon mean free path, which is vt, and t is the average time between collisions. Thus, the thermal conductivity of solid material is reduced by lowering the group velocity or shortening the mean-free path. The mean free path of a phonon decreases as temperature increases. A shorter mean-free path is the main reason why many superior thermoelectric materials possess the highest ZT values. [5,18]

At high temperatures (500–1000K), phonons are scattered by point and line defects because l scales similarly to lattice spacing. Phonon scattering of defects such as interfaces and grain and sub-grain boundaries generally reduce thermal conductivity at temperatures near ambient. At 300K, the dominant phonon wavelength is 1 nm. [20] Large unit cells of heavy elements and alloys such as SiGe also promote phonon scattering. Besides phonon-phonon scattering within a crystal, the introduction of point and line defects enhances scattering. Interfaces such as grain boundaries, inclusions, and pores can also scatter phonons at microscopic levels.

6.2 Thermoelectric Theories and Limitations

6.2.1 Best Thermoelectric Materials

The atomic electron bandgap of thermoelectric materials has a direct influence on thermoelectric couple performance. Chasmar and Stratton [21], followed by Mahan and Sofo [22], developed theories on improving thermoelectric materials. They derived a single parameter B to maximize ZT,

$$B \propto T^{5/2} \frac{m^{*1/2}\tau}{\kappa_L} \tag{6.11}$$

where τ is electron scattering time, and κ_L is the lattice thermal conductivity. The effective mass is proportional to the bandgap energy when considering phonon scattering from impurities and optical and acoustic phonons. Thus, the optimum bandgap for a thermoelectric material is between 6 and 10 k_BT.

In their publication, *The Best Thermoelectric*, Mahan and Sofo [23] used electronic band theory and a known lattice thermal conductivity to formulate the best electronic structure with the highest figure of merit; see equations (6.12–6.16). Using solutions for Boltzmann equations for transport coefficients, σ, S, and κ,

$$\sigma = e^2 \int_{-\infty}^{+\infty} d\varepsilon \left(-\frac{\partial f_0}{\partial \varepsilon} \right) \Sigma(\varepsilon) \tag{6.12}$$

$$T\sigma S = e \int_{-\infty}^{+\infty} d\varepsilon \left(-\frac{\partial f_0}{\partial \varepsilon} \right) \Sigma(\varepsilon)(\varepsilon - \mu) \tag{6.13}$$

$$T\kappa_0 = \int_{-\infty}^{+\infty} d\varepsilon \left(-\frac{\partial f_0}{\partial \varepsilon} \right) \Sigma(\varepsilon)(\varepsilon - \mu)^2 \tag{6.14}$$

where $\Sigma(\varepsilon)$ is a transport distribution function:

$$\Sigma(\varepsilon) = \Sigma_{\bar{k}} v_x(k) \tau(k)^2 \sigma(\varepsilon - \varepsilon(k)) \tag{6.15}$$

They defined κ_0 as the electronic contribution of thermal conductivity when the electrochemical potential gradient inside the sample is zero. Using the relationship,

$$\kappa_\varepsilon = \kappa_0 - T\sigma S \tag{6.16}$$

we can find the $\Sigma(\varepsilon)$ that will maximize ZT. They conclude that only a Dirac delta distribution can satisfy the conditions for a maximum ZT, which means a delta function in the density of state (DOS) is preferred. Mahan-Sofo recognized this kind of energy distribution does not exist in real-world material. However, their work provided explanations on some existing good thermoelectric materials and pointed a path to search for materials with narrow energy distribution of carriers with high velocity along the direction of the applied field. They pointed out the d- and f- bands are likely to have a DOS that resembles the delta function.

Deng et al. [24] followed this path to search for materials with a narrow DOS near the band edge. They identified poly (tetrathienoanthracene) (PTTA) from a few covalent organic frameworks. The PF of these materials ranged from 14.9–21.9 μW cm^{-1}·K^{-2}. For high-performance thermoelectrics such as SiGe, skutterudite, and PbTe, the PF is in the range of 20–50 μW cm^{-1}·K^{-2}. In the reviews by Shakouri [7] and Yang [10], the Seebeck coefficient for differential conductivity, known as the Mott formula, is expressed by two equivalent forms in the following equation (6.17):

$$S = \frac{\pi^2}{3} \frac{k_B^2 T}{q} \frac{d\ln[\sigma(E)]}{dE}\bigg|_{E=E_F} \quad or \quad \frac{\pi^2}{3} \frac{k_B^2 T}{q} \left(\frac{1}{n} \frac{dn(E)}{dE} + \frac{1}{\mu} \frac{d\mu(E)}{d\mu} \right)_{E=E_F} \tag{6.17}$$

where μ represents the mobilities of carriers. They also concluded that to maximize the Seebeck coefficient, the material's electrical conductivity should be as high as possible to achieve high degrees of asymmetry in the Fermi energy. The trade-off and balance between the Seebeck coefficient and electrical conductivity mean that the maximum PF highly depends on the shape of the DOS curve and electron group velocity (related to effective mass). Materials with high electron-effective mass and DOS with multiple valleys have the potential to have high ZT. However, these materials usually have low electron mobility, which is not universal. For example, SiGe is a proven thermoelectric material with high electron mobility and low-effective mass. Materials like SiGe do not follow this relationship and often have a slight electronegativity difference between atoms. The latter expression in equation (6.17) shows that increased sharpness (first derivative) of carrier concentration n and mobility μ will increase the Seebeck coefficient. These relationships are used for band engineering and carrier scattering. Goldsmid recognized that fixed carrier concentration and energy degenerate band edges enhance the Seebeck coefficient. [25] Carrier scattering can be manipulated by electron-electron, electron-phonon, and impurity scatterings.

The insights of band structure and the influence of DOS on PF prompted work in "band engineering." Hicks and Dresselhaus [26,27] also pointed out that sharp features designed into the DOS of a quantum structure also enhance PF. Similar strategies were proposed by others. [28,29] While not successful in producing practical thermoelectric devices, these strategies further enhanced beliefs on scattering by dopants, impurities, and other defects in bulk thermoelectric materials. A good example is PbTe, where Chen [17] revealed non-parabolic edge states at the energy pockets near the band edge. The doping of Tl in PbTe [30] is an example of band engineering resulting in significantly improved PF and ZT.

6.2.2 Imbalanced Thermoelectric Legs

One interesting observation regarding the relationship between thermoelectric performance and materials selection is the imbalance between n-type and p-type materials. Good performance of one thermoelectric material does not automatically translate to an equivalent performance of the opposite type of thermoelectric material in the same crystal structure. There are plenty of examples in RTG-related materials. For example, the n-type PbTe does not have an equally matching p-type leg. The Multi-Mission RTG (MMRTG) does not have a matching p-type PbTe [31]; instead, the p-leg is a segmented architecture of PbSnTe and TAGS, where TAGS is an alloy of tellurium, germanium, antimony, and silver. [32-34] In SiGe, the ZT of the n-type leg is higher than the p-type leg. [35,36] Similar examples exist in thermoelectric skutterudite systems. [37-41] Some traditional thermoelectric systems with imbalanced n-type and p-type legs have seen significant improvement in recent years. New materials such as 14-1-11 Zintl compounds [42], lanthanum tellurides [43], silicides [44-46], and tetrahedrites [47] have only one material type with a high ZT. Thus, different thermoelectric material legs are used with these new materials for high-performance systems.

This imbalanced phenomenon is closely related to the underlying physics tied to band structure and the sharpness of DOS at the band edges. Chemical composition and crystal structure determine the basic band structure of a particular undoped material. The shape of the DOS and asymmetry near the band edge, with respect to E_F, and optimized for one type of dopant rarely favors the opposite type of thermoelectric material. For these material systems, an advantage that leads to a higher PF of one type of material can make it impossible to achieve the same PF just by doping the same material. The imbalance is primarily because of electron transport. For the oldest thermoelectric material, bismuth telluride, Bi_2Te_3, the n-type material is $(Bi_xSb_{1-x})_2Te_3$, and the p-type material is $Bi_2(Se_xTe_{1-x})_3$. The x value in n-type material is lower than 0.25, making the material mostly Sb_2Te_3. Although closely related to the electrical conductivity of the materials, thermal conductivity plays a less critical role in imbalanced thermoelectric legs.

Understanding the imbalance of thermoelectric materials and the underlying physics is essential in optimizing the performance of thermoelectric devices. On the one hand, the band analysis can help us understand why a good match in *ZT* could not happen by doping the same base material in specific single-legged thermoelectric materials. The best alternative might be a different material. On the other hand, it is sometimes helpful to highlight the problem of dopant influence. Thus, band engineering and carrier scattering mechanisms can improve the *ZT* of both legs.

6.3 Thermal Conductivity and Phonon Scattering

As shown by equation 6.4, the PF must be as large as possible to maximize *ZT*, while thermal conductivity should be minimized. Thus, in addition to using band engineering to maximize the PF, research has focused on increasing phonon scattering and lowering lattice thermal conductivity.

Skutterudites, which will be discussed in a later chapter, are thermoelectric materials based on $CoSb_3$. These materials may potentially replace Pb/TAGS in the MMRTG because of their higher operating temperature range of 500–600°C. [48] The crystal structure of skutterudites is shown by Wang et al. [49] in Figure 6.5.

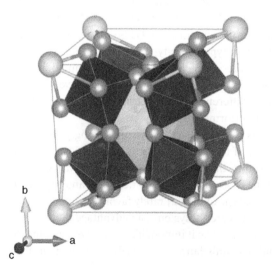

Figure 6.5 Crystal structure of skutterudite, with the filler atoms in cyan, the Sb atoms in tan, and the Co atoms in the dark blue octahedra. The large cages are centered on the corners and in the center of the unit cell. [49]

Interestingly, the cage-like voids in the skutterudite structure accommodate large diameter dopants such as Ba and Yb. [50] These cages have been described as rattlers, which scatter phonons to reduce thermal conductivity. The term phonon-glass electron-crystal (PGEC) refers to materials like skutterudites and clathrates. [51] For example, Xun et al. [52] used double or triple dopants instead of a single dopant [53] to fill up the cages, leading to further thermal conductivity reductions. In addition, the scattering of phonons off La, Ba, and Yb atoms are sufficiently different to cause further reductions of thermal conductivity, thus achieving higher ZT.

6.3.1 Highlights of SiGe

We highlight SiGe, which is a material system still being used by NASA, to summarize the importance of the underlying physics of thermoelectric materials. Compared to the recent versions of this material system, the ZT of heritage SiGe is the highest available to replicate at production scale. SiGe is an excellent example of how understanding electron and phonon transport in solids guided the discovery and development of SiGe for the Multi-Hundred Watt (MHW) RTG. The pioneering transport work by Ioffe [54] in the 1950s influenced the development of SiGe by Vining and Fleurial. [55]

Additionally, work published in 1954 by Herrin [6] suggested that phonon scattering affected thermoelectric power. Slack and Hussain [35] presented the concept of "phonon drag" and proposed methods to optimize the efficiency of SiGe in 1991. They considered all the methods to maximize ZT in their analyses, including doping, electron, phonon scattering, effective mass, lowering thermal conductivity, etc. They concluded that the efficiency of SiGe could be at least 12.7%, with an ultimate value of 23.4%. More recent work by Shen et al. [56] showed that alloying silicon and germanium made the mixed crystals very effective at scattering phonons while not degrading electronic transport. These fundamental research efforts by generations of researchers are critical to the success of thermoelectric materials, RTGs, and future space missions.

Subsequent chapters will discuss advanced thermoelectric materials and technologies for RTGs in greater detail.

References

1 Seebeck, T.J. (1822). "Über den Magnetismus der galvanischen Kette". Abhadlungen der Physikalischen Klasse der Königlisch-Preußsischen. *Akademie der Wissenschaften aus den Jahren* 1820–1821: 289–346.

2 Peltier (1834). "Nouvelles expériences sur la caloricité des courants électrique" [New experiments on the heat effects of electric currents]. *Annales de Chimie et de Physique (in French).* 56: 371–386.

3 Thomson, W. (1851). On a mechanical theory of thermoelectric currents. In: *Proceedings of the Royal Society of Edinburgh*, 91–98. doi:10.1017/S0370164600027310.

4 Jeffrey Snyder, G. and Toberer, E.S. (2008). Complex thermoelectric materials. *Nat. Mater.* 7: 105–114.

5 Zevalkink, A., Smiadak, D.M. et al. (2018). A practical field guide to thermoelectrics: Fundamentals, synthesis, and characterization. *Appl. Phys. Rev.* 5: 021303.

6 Herring, C. (1954). Theory of the Thermoelectric Power of Semiconductors. *Phy. Rev.* 96 (5): 1163 1187.

7 Shakouri, A. (2011). Recent Developments in Semiconductor Thermoelectric Physics and Materials. *Annu. Rev. Matter. Red* 41: 339–431.

8 Tian, Z., Lee, S., and Chen, G. (2014). A Comprehensive Review of Heat Transfer in Thermoelectric Materials and Devices. *Annu. Rev. Heat Transf.*, https://doi.org/10.1615/AnnualRevHeatTransfer.2014006932 425–483.

9 Mahan, G.D. (2016). Introduction to Thermoelectrics. *APL Mater.* 4: 104806.

10 Yang, J. (2016). On the tuning of electrical and thermal transport in thermoelectrics: an integrated theory–experiment perspective. *Npj Comput. Mater.* 2: 10015.

11 Bube, R.H. (1987). *Electronics in Solids – An Introductory Survey*, 2e. San Diego CA: Academic Press Inc.

12 Rowe, D.M. (ed.) (2005). *CRC Handbook of Thermoelectrics: Macro to Nano.* Boca Raton: CRC.

13 Chen, G., Dresselhaus, M.S., Dresselhaus, G. et al. (2003). Recent developments in thermoelectric materials. *Int. Mater. Rev.* 48: 45–66.

14 Tritt, T.M. (ed.) (2001). *Recent Trends in Thermoelectric Materials Research.* San Diego: Academic.

15 Ben, F. et al. (2002). Electronic structure and optical properties of Si1-xGex alloys. *Physica* B.322: 225–235.

16 Madsen, G.K.H. and Singh, D.J.B.T.P. (2006). A code for calculating band-structure dependent quantities. *Comput. Phys. Commun.* 175: 67–71.

17 Chen, X., Parker, D.D., and Singh, D.J. (2013). Importance of non-parabolic band effects in the thermoelectric properties of semiconductors. *Sci. Rep.* 3 (1): 3168.

18 Chung, J.D., McGaughey, A.J.H., and Kaviany, M. (2004). Role of Phonon Dispersion in Lattice Thermal Conductivity Modeling. *Trans. ASME* 126: 376–380.

19 Wei, S. and Zhou, M.Y. (1994). Phonon dispersions of silicon and germanium from first-principles calculations. *Phys. Rev. B* 50: 2221.

20 Bourgeois, O., Tainoff, D. et al. (2016). Reduction of phonon mean free path: From low-temperature physics to room temperature applications in thermoelectricity. *C. R. Phys.* 17 (10): 1154–1160.

21 Chasmar, R.P. and Stratton, R. (1959). The thermoelectric figure of merit and its relation to thermoelectric generators. *J. Electron. Control* 7 (1): 52–72.

22 Sofo, J.O. and Mahan, G.D. (1994). Optimum band gap of a thermoelectric material. *Phys. Rev. B* 49: 4565–4570.

23 Mahan, G.D. and Sofo, J.O. (1996). The Best Thermoelectric. *Proc. Natl Acad. Sci. USA* 93: 7436–7439.

24 Deng, T. et al. (2020). Beyond the Mahan–Sofo best thermoelectric strategy: high thermoelectric performance from directional p-conjugation in two-dimensional poly(tetrathienoanthracene). *J. Mater. Chem. A* 8: p4257.

25 Goldsmid, H.J. (2010). *Introduction to Thermoelectricity*. Springer.

26 Hicks, L.D. and Dresselhaus, M.S. (1993). Effect of quantum well structure on thermoelectric figure of merit. *Phys. Rev. B* 47 (19): 727–731.

27 Dresselhaus, M.S. et al. (2007). New directions for low-dimensional thermoelectric materials. *Adv. Mater.* 19: 1043–1053.

28 Harman, T.C., Taylor, P.J., Walsh, M.P., and LaForge, B.E. (2002). Quantum Dot Superlattice Thermoelectric Materials and Devices. *Science* 297 (5590): 2229–2232.

29 Venkatasubramanian, R. and Siivola, E. (2001). Thomas Colpitts and Brooks O'Quinn, Thin-film thermoelectric devices with high room-temperature figures of merit. *Nature* 413: 597–602.

30 Heremans, J.P. et al. (2008). Enhancement of thermoelectric efficiency in PbTe by distortion of the electronic density of states. *Science* 321: 554–557.

31 Pei, Y., Lensch-Falk, J., Toberer, E.S. et al. (2010). High Thermoelectric Performance in PbTe Due to Large Nanoscale Ag2Te Precipitates and La Doping. *Adv. Funct. Mater.* 21 (2): 241–249.

32 Schock, A. (1992). RTGs using PbTe/TAGS thermoelectric elements for Mars environmental survey (MESUR) mission. Fairchild Space Report *FSC-ESD-217-92-509*, 1–12.

33 Nolas, G.S., Sharp, J., and Goldsmid, H.J. (2001). *Thermoelectrics: Basic Principles and New Materials Developments*. New York: Springer.

34 Skrabeck, E. and Trimmer, D.S. (1995). *CRC Handbook of Thermoelectrics* (ed. D.M. Rowe), 267. Boca Raton, FL: CRC Press.

35 Slack, G.A. and Hussain, M.A. (1991). The maximum possible conversion efficiency of silicon – germanium thermoelectric generators. *J. Appl. Phys.* 70: 2694.

36 Minnich, A.J. et al. (2009). Modeling study of thermoelectric SiGe nanocomposites. *Phys. Rev. B* 80 (15): 155327.

37 Caillat, T., Borshchevsky, A., and Fleurial, J.-P. (1996). Properties of single crystalline semiconducting CoSb3. *J. Appl. Phys.* 80 (8): 4442–4449.

38 Morelli, D.T., Caillat, T., Fleurial, J.-P. et al. (1995). Low- temperature transport properties of p-type CoSb3. *Phys. Rev. B* 51 (15): 9622.

39 Nolas, G.S., Slack, G.A., and Schujman, S.B. (2000). *Semicond. Semimet.* 69: 255.

40 Sales, B.C., Mandrus, D., and Williams, R.K. (1996). *Science* 272: 1325.

41 Nolas, G.S., Kaeser, M., Littleton, R.T., and Tritt, T.M. (2000). *Appl. Phys. Lett.* 77 (12): 1855.

42 Fisher, I.R. et al. (2000). Yb14ZnSb11: Charge Balance in Zintl Compounds as a Route to Intermediate Yb Valence. *Phys. Rev. Lett.* 85(5: 1120–1123.

43 May, A.F., Fleurial, J.-P., and Jeffrey Snyder, G. (2008). Thermoelectric performance of lanthanum telluride produced via mechanical alloying. *Phys. Rev. B* 78: 125205.

44 Kim, S.G., Mazin, I.I., and Singh, D.J. (1998). *Phys. Rev. B* 57: 6199.

45 Zaitsev, V.K., Fedorov, M.I., Gurieva, E.A. et al. (2006). *Phys. Rev. B* 74: 045207.

46 Fukano, M., Iida, T., Makino, K. et al. (2007). *MRS Proc.* 1044: https://doi. org/10.1557/PROC-1044-U06-13.

47 Xu, L. and Morelli, D.T. (2013). Natural mineral tetrahedrite as a direct source of thermoelectric materials. *Phys. Chem. Chem. Phys.* 15: 5762–5766.

48 Holgate, T.C., Bennett, R., Renomeron, L. et al. (2019). Analysis of Raw Materials Sourcing and the Implications for the Performance of Skutterudite Couples in Multi-Mission Radioisotope Thermoelectric Generators. *J. Electron. Mater.* 48 (11): 7526–7632.

49 Wang, H., Kirkham, M.J., Watkins, T.R. et al. (2016). Neutron and X-ray powder diffraction study of skutterudite thermoelectrics. *Powder Diffract.* 31 (1): 16–22.

50 Salvador, J.R., Yang, J., Wang, H., and Shi, X. (2010). Double-filled skutterudites of the type YbxCayCo4Sb12: Synthesis and properties. *J. Appl. Phys.* 107 (4): 043705.

51 Slack, G.A. (1995). *CRC Handbook of Thermoelectrics* (ed. D.M. Rowe) Ch. 9. CRC Press.

52 Shi, X., Yang, J., Salvador, J.R. et al. (2011). Multiple-filled skutterudites: high thermoelectric figure of merit through separately optimizing electrical and thermal transports. *J. Am. Chem. Soc.* 133 (20): 7837–7846.

53 Salvador, J.R., Yang, J., Shi, X. et al. (2009). Transport and mechanical properties of Yb-filled skutterudites. *Philos. Mag.* 89 (19): 1517–1534.

54 Ioffe, A.F. (1957). *Semiconductor thermoelements, and Thermoelectric cooling,* Reviewed and supplemented for the English edition. (Tran. ed. A. Gelbtuch). Infosearch.

55 Vining, C.B. and Fleurial, J.-P. (1993). Silicon-Germanium: an overview of recent developments. Tenth Symposium on Space Nuclear Power and Propulsion, Special Volume Albuquerque, New Mexico (10–14 January 1993).

56 Shen, L. et al. (2021). The effect of vacancy defects on the conductive properties of SiGe. *Phys. Lett. A* 386: 126993.

7

End-to-End Assembly and Pre-flight Operations for RTGs

Shad E. Davis

Idaho National Laboratory, Idaho Falls, Idaho

7.1 GPHS Assembly

The core of a Radioisotope Thermoelectric Generator (RTG) is a radioisotopic heat source. The US uses General Purpose Heat Sources (GPHSs) as the essential building blocks of a radioisotopic heat source for RTGs. Early design goals and attributes of the GPHS included modularity to allow for flexibility and compatibility with static and dynamic systems, high power density to produce 165.5 thermal watts per kg (75 W_{th}/lb), improved safety backed by extensive safety testing, and lower cost. [1]

The first iteration of a GPHS, designated Step-0, was used to fuel the GPHS-RTGs used by the Galileo, Ulysses, and Cassini spacecraft; see Figure 7.1.

A small but critical design modification that added a "webbing" between the cavities for the graphite impact shells (GISs) enhanced the Step-0 GPHS. This increased the strength of the aeroshell and earned it the name Step-1 GPHS. The GPHS-RTG for the Pluto New Horizons utilized the Step-1 GPHSs. Today, the Step-2 GPHS is in use. The thickness of the top and bottom surfaces were increased to strengthen those elements and add margin against ablation induced by aerodynamic heating experienced during an inadvertant re-entry scenario. Step-2 GPHS modules were used to fuel the Multi-Mission Radioisotope Thermoelectric Generator (MMRTG) aboard the Curiosity and Perseverance rovers.

A Step-2 GPHS is a module or block that measures approximately 10 cm × 10 cm × 5.8 cm (4"×4"×2.3") and weighs around 1.6 kg (3.5 lbs) each. Each GPHS module contains a thermal power output of around 250 watts once fueled. Each module houses four plutonium-238 oxide fuel pellets or clads.

The Technology of Discovery: Radioisotope Thermoelectric Generators and Thermoelectric Technologies for Space Exploration, First Edition. Edited by David Friedrich Woerner.

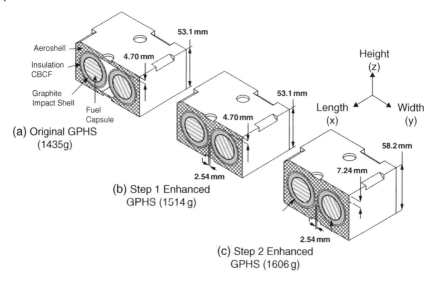

Figure 7.1 Lipinski, Ronald J., Hensen, Danielle L. "Criticality Calculations for Step-2 GPHS Modules" 2008.). (a-c) Cutaway View of Step-0, Step-1, & Step-2 GPHS Modules (Model Figure: Lipinski, Ronald J., Hensen, Danielle L. "Criticality Calculations for Step-2 GPHS Modules" 2008). Credit: DOE/Sandia National Laboratory.

Plutonium-238 is the heat source material used today because of its characteristics during decay. ^{238}Pu is an alpha emitter, meaning it releases the nucleus of a helium atom during disintegration. This alpha particle is large enough that the fuel absorbs all the energy of disintegration, creating heat. The half-life of ^{238}Pu is ideal because it is active enough to produce heat with reasonable quantities of material, but long enough to support decades long missions with an acceptable loss of thermal inventory. Helium gas is the byproduct of this decay.

Each plutonium-oxide pellet is encompassed within vented, welded clads made of a high-strength iridium alloy. Iridium metal provides high strength, good ductility, and good oxygen resistance, all of which provide favorable characteristics that limit release of radioisotope contamination following an unlikely launch accident. The iridium alloy, fabricated at Oak Ridge National Laboratory (ORNL) in Tennessee, is compatible with the plutonium-238 oxide fuel. Contact with plutonium fuel does not change its physical and mechanical properties. Each Fuel Clad (FC) contains a porous media filter, or frit vent, designed to release the helium yet, does not allow the release of plutonium particulates from the FC. This intentional leak prevents an FC from a pressure buildup that could cause cracking, or worse, a breach,of an iridium clad, allowing respirable particles to escape, thus harming individuals following an accident.

Figure 7.2 shows a detailed cutaway view of the components that comprise a GPHS. Two FCs are inserted into a cylindrical container made of Fine Weave

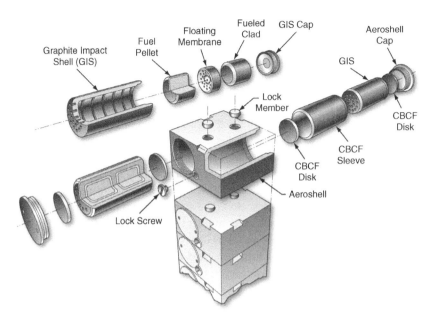

Figure 7.2 Cutaway and Exploded View of Step-0 GPHS Module and Stack. Model Image: INL RPS Program.

Pierced Fabric (FWPF) graphite composite known as a Graphite Impact Shells (GISs). GISs provide a layer of protection against impacts and shrapnel that could result from a launch pad accident or re-entry accident. A floating membrane of FWPF separates FCs inside a GIS. A thermal insulator sleeve made from Carbon Bonded Carbon Fiber (CBCF), fabricated at ORNL, surrounds each GIS. This limits the maximum temperature rise the FCs experience during an inadvertent re-entry and maintains the iridium clad temperature above the brittle/ductile transition temperature during normal operations, reducing the potential of a breach upon impact. The GPHS aeroshell, also made of FWPF, is loaded with two GIS/CBCF sleeve assemblies, one in each cavity. Each cavity is capped and each cap is locked into place with a screw. A FWPF aeroshell serves as the primary structure of a GPHS and provides protection against impacts on hard surfaces, re-entry heating, and launch pad accidents that could cause overpressure insults and impacts to a GPHS.

Each component of a GPHS delivers specific functions. Key among them is that a GPHS and its components facilitate heat transfer from fuel to thermoelectric couples in an RTG. In addition, each GPHS and its internal components provide protective functions to contain and prevent release of the plutonium-238 fuel, to the maximum extent possible, should an accident occur during any phase of the mission, including ground handling, transportation, and launch and ascent of an RTG. Potential credible accident scenarios include explosion of a launch vehicle on a launch pad, launch accidents resulting in atmospheric re-entry, subsequent impact, and post-impact fires.

Fuels clads for GPHSs are assembled at the Los Alamos National Laboratory (LANL) in New Mexico by pressing an oxide pellet to shape and then welding the pellets into an iridium fuel clad. Each clad has a frit vent that is sealed with a welded cap of iridium so that a helium leak test can verify the integrity of the welded assembly. The caps on the vents will be removed later, at the Idaho National Laboratory. Thermal calorimetry is performed to measure the heat inventory of every FC. The FCs are shipped in casks and shipped to Idaho National Laboratory (INL) for assembly into GPHSs.

The Department of Transportation (DOT) has made it clear radioactive materials may be transported in 9516 shipping casks. Transportation is limited to fewer than 500 watts thermal of plutonium dioxide heat source material in any solid-state form (e.g., powder, pellets, granules, etc.). The stainless steel Containment Vessel (CV) inside the 9516 shipping package container is leak-tight and fabricated to US national standards for boiler pressure vessel codes. GPHS fuel clads can be configured inside the CVs in two ways. Either method uses stainless steel product cans (PCs) that contain the fuel clads inside a CV. [2] Several technical security requirements are enforced to protect nuclear operators, co-located area workers, and the public from any potential release of plutonium-238 oxide and in order for INL to accept any GPHS fuel clads into the facility. A complete certificate of inspection containing the data documenting fabrication of each GPHS fuel clad is also a technical security requirement that is enforced before clads are allowed into INL.

Several processes are exercised before GPHS module assembly begins. Each component and its serial number are documented in a build sheet. Each build sheet identifies specific fuel clads destined for a specific GPHS to set the thermal inventory of the RTG heat source stack at beginning of mission. Managing the total thermal inventory in a heat source stack is essential for meeting an RTG's power output requirement.

Pre-assembly processes for the FWPF components perform precise measurements of the soon to be assembled GPHS. Most notably, the overall height of each GIS is determined, so that the tapered gap between the FWPF aeroshell cap and FWPF GIS cap can be set to ensure a thermal pathway between the GPHS aeroshell and its components at nominal heat source operating temperatures.

Extensive studies, analyses, and tests led to a design that optimized the room temperature gaps within the FWPF aeroshell's insulating and impact components and thus established "the gap" that is still measured and accounted for in pre-assembly today. The process for measuring and verifying compliance with the gap specification and follow on repeatable establishment of that gap has improved over the years. Each gap measurement now occurs during GPHS module fueling and assembly and as discussed later in this chapter.

Every end cap is threaded into the aeroshell, therefore each aeroshell cavity can be used for any GIS. The GIS measurement is used to set an adjustable simulated

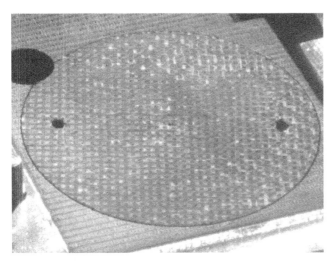

Figure 7.3 GPHS aeroshell cap with lune form made in upper left corner. Credit: US Department of Energy..

GIS at the exact height, the GIS height, plus the design gap. The fixture is loaded into the aeroshell and the threaded end cap is torqued to specifications, creating a unique assembly height for the specified GIS. A jig grinder, dedicated for one purpose, is used to cut a crescent-shaped form (known as a lune), shown in Figure 7.3, into the FWPF aeroshell and a cap at the exact location where a locking screw will be installed, ensuring the required precise assembly of a GPHS module. The jig grinder is capable of three cutting motions at a cutting speed of 20,000 revolutions per minute. Figure 7.4 shows the jig grinder used for making the lune form cut in the FWPF aeroshell.

The cutting process is smooth. A US five cent coin stood on edge on the head of the jig grinder while the machine runs at full speed, remains upright. Thermal expansion of the grinder could cause grindings of the crescent-shaped cut into an aeroshell cap to be unfit for spaceflight. The smallest mistake can cause scrapping of an entire part, a part that may have no replacement in inventory. The head stock bearings of the grinder are heated at all times, even when the machine is not in use to prevent this issue. Each lune cut matches an aeroshell body with a GIS and an aeroshell cap. An accurate lune cut ensures the size of the critical gap.

Pre-assembly operations also include a vacuum bake-out process. The graphitic components of each GPHS module are baked out in a high temperature vacuum oven to remove moisture, oxygen, and other contaminants that may be present. These graphitic components are never exposed to air after they are baked. All handling of the graphitic components is done in inert gloveboxes. Transfers between gloveboxes are conducted with the components inside

Figure 7.4 GPHS module set-up on the jig grinder before performing an aeroshell end cap lune cut. Credit: US Department of Energy.

hermetically sealed cans filled with inert gas, which maintains the baked out and "clean" condition. The graphitic components of a GPHS module are transferred to the module assembly glovebox, where they are staged before integration with fuel clads.

INL receives the fuel shipment and unpacks each 9516 package by removing the CVs from the 9516 package and cutting open the CVs and their liners to retrieve the PCs. A machined groove on the outside of each CV marks the cut location. Each cut is performed while a chuck rotates a CV in a ventilated fume hood. The hood provides a protective barrier against unlikely, but potential release of radio-active contamination during the cutting operation. The PCs are retrieved from the CV liner and then transferred to a glovebox where the FC can be removed. Figure 7.5 shows opened PCs and the fuel clads retrieved.

One end of each PC is magnetized, and this is used to ensure proper orientation when inserting a PC into a cutting fixture. In addition, the PCs have pre-machined grooves similar to the CV to show the location for a cut.

The build of each GPHS is planned out before assembly. A fuel clad is assigned to be assembled together with three other fuel clads along with the serialized components for a GPHS. This planning is documented in a GPHS module and GIS build sheet. The build sheet is a quick and easy visual reference of planned placement of fuel clads in a GPHS and subsequent ordering of stacked GPHSs for insertion into an RTG.

Figure 7.5 Two GPHS fuel clads retrieved from a cut-open PC and placed in the tray on the right. Credit: US Department of Energy.

Each GPHS assembly occurs under an inert atmosphere in a glovebox. That atmosphere is continuously scrubbed to remove oxygen and moisture down to a few parts per million. Controlling moisture and oxygen is important in protecting the graphite components of a GPHS from oxidation that could degrade and subsequently fail in a launch accident. A secondary purpose for maintaining a low oxygen environment is to preserve the fuel pellet chemistry related to the oxygen isotope of the plutonium dioxide fuel.

Each fuel clad arriving from LANL is filled with helium because they were assembled in a helium-rich atmosphere and because the natural decay of the plutonium-238 fuel produces helium. Therefore, helium can be detected if a clad has a leak. Each clad is leak checked using a helium leak detector. Fuel clads that pass the leak check are near ready for assembly into their GISs.

Each GPHS fuel clad includes a closed frit vent. The vent is made from a porous, sintered iridium disk that is welded to one end of each GPHS fuel clad. The frit vent should release helium at a sufficient rate to prevent overpressurization of the fuel clad that could cause the fuel clad to rupture. An iridium barrier protects the frit vent when it arrives at INL and the barrier is called a decontamination cover. Frit vent covers are sheared open using a lance prior to GIS assembly.

Each GIS contains two FCs that are separated by a floating membrane. GIS caps secure the contents of each GIS and are torqued into place. Several critical measurements are made on each assembled GIS. INL has adopted the practice of assembling GISs well before placing them into GPHSs. This technique, of advanced assembly of GISs, opened the opportunity to improve how assembly campaigns are conducted regarding the gap measurement between the GIS and aeroshell cap, as well as the thermal inventory of each GPHS module. Once measurements are taken to determine the length of a particular fueled GIS, for example, the measurements can then be used to set the length of an adjustable simulated GIS. Simulated GISs are used to set gaps and inform the location of a lune form cut in the aeroshell cap. Two fueled GISs are then inserted into the cavities of a GPHS module, one side at a time. Figure 7.6 shows a GIS being inserted into the GPHS module cavity.

Extreme care is taken when inserting the GISs into the sleeves inside the GPHS cavities to avoid damaging the fragile sleeves. Aeroshell caps are threaded into the opening of each GPHS cavity until the crescent shaped lune cut in the aeroshell body and cap align. A lock screw member is threaded and torqued into the aeroshell body and cap interface to secure the aeroshell cap in place. The fueling of a GPHS is now complete, and a GPHS is ready for the Module Reduction Monitoring (MRM) process.

The MRM process reduces the oxygen potential by reducing the stoichiometry of the fuel. Oxygen evolving from the fuel reacts with the carbon of the graphitic components, creating carbon monoxide (CO), an undesirable compound in an RTG. The reduction occurs when oxygen from the fuel evolves and diffuses through the fuel clad frit vent where it reacts with the carbon in the graphite of the GPHS reacting to form CO. CO migrates through the frit vent and reacts with the oxygen from the fuel, which reduces the fuel by creating carbon dioxide (CO_2). CO_2 passes out of the

Figure 7.6 A fueled GIS being lowered into a GPHS cavity using a vacuum lift tool and assembly fixture inside an INL glovebox. Credit: US Department of Energy.

clad through the frit vent and contacts the graphitics, creating two CO molecules. This cyclic reaction continues until the gas byproducts of CO and CO_2 achieve an acceptable level believed to be a substoichiometric value approaching 1.98. [3]

7.2 RTG Fueling and Testing

Illustrative examples of operations at INL are described and current practices conveyed, yet generic processes are described to give readers a general understanding of RTG fueling and testing practices. Handling processes used for the MMRTG are described as needed.

Most flight generators arrive at INL configured with an electrical heater months before the time when fueling and assembly operations begin. Figure 7.7 shows the electrically heated flight unit received at INL.

Figure 7.7 An ETG received at INL. This ETG became an RTG once fueled and was used on the Mars 2020 Perseverance Rover that launched on 30 July 2020. Credit: US Department of Energy.

Each unit arrives in a custom shipping container that protects the thermoelectric generator. The shipping containers protect the hardware from excessive shock and vibration during transportation. A nitrogen atmosphere pressurizes each container, and each container carries moisture absorbing desiccants to protect the hardware's surfaces from gaseous and condensing moisture. The electrical heaters simulate a generator's fuel. These electrically heated thermoelectric generators go by the acronym ETG. A series of electrical tests and checks conducted upon arrival at INL verify electrical resistance and grounding of the thermoelectric circuit in an ETG and those checks verify the generator is undamaged. It is critical for quality control that no electrical shorts developed because of shipping and that resistance values do not show a negative change in the overall condition of internal grounding connections. The health of internal grounding connections is important because it shows to mission integrators that the intended means of bonding, shielding, and grounding an RTG with a spacecraft will work. A thorough visual inspection of the exterior of each ETG scans for any damage or visible defects. Findings are mapped and documented to distinguish those found from any damage or defects incurred at INL. Each ETG has an emissive coating on the outside surfaces of the generator housing and the fins that are attached to the outside of the housing. The emissive coating plays a critical role in transferring heat from the generator.

Fueling and assembling an RTG combines an ETG with fueled GPHSs. Much of the process is slow and meticulous and requires a great deal of skill, patience, and attention to detail. The fueling and assembly process occurs in INL's Inert Atmosphere Assembly Chamber (IAAC). The IAAC is a hot cell that was built with sophisticated remote manipulators and glove ports on the side of the chamber that allow skilled operators to conduct the entire fueling and assembly process at a safe distance from the fuel, protected from radiation.

An ETG mated with its electrical ground support equipment is moved into the IAAC. The ground support equipment monitors the generator from outside the IAAC, and the ground support equipment performs multiple functions, such as providing an electrical load to the generator output circuit, reads, and records temperatures from the temperature sensors inside and outside of the generator.

Each ETG is stripped of its electrical heater and any ancillary hardware that is not needed for flight. The cavity that remains, as shown in Figure 7.8, is inspected and cleaned. The ETG is ready to be fueled and become an RTG. Operators fill the IAAC with an inert atmosphere and the IAAC's environmental control system maintains a low level of oxygen and moisture.

The fueling process begins by introducing GPHS modules encapsulated in MRM cannisters into a glovebox that is connected to one end of the IAAC. Cuts are made to open the MRM cannisters, and the GPHSs removed. A GPHS module inside a MRM canister is shown in Figure 7.9.

Figure 7.8 An MMRTG's cavity following removal of the electrical heater and associated components. Credit: US Department of Energy.

Figure 7.9 A fueled GPHS module inside the MRM canister being prepared for fueling the RTG. Credit: US Department of Energy.

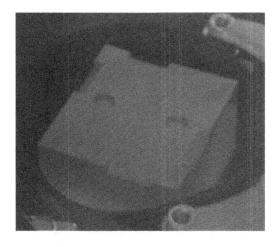

The GPHSs are cleaned, inspected, their serial numbers verified, and moved into the IAAC to create a stack of GPHSs. Initially, a GPHS stack begins by placing a GPHS on an interface plate that has features that allow an entire GPHS stack to be lifted and lowered into an RTG. An interface plate aids in aligning a GPHS stack in the liner cavity of a generator. Getters installed on the plate remove oxygen and nitrogen that ingresses into the RTG on Earth. The order of the GPHSs in a stack is pre-determined, and each GPHS is locked into place by two lock members. The lock members keep the stack aligned during the fueling process, subsequent testing, and mission loads. The clocking of each GPHS module is important as the machined bevels are a design feature that supports launch accident re-entry analysis and modeling. The stack is ready for insertion into an RTG.

A lift fixture engages the features of the interface plate and an entire stack is craned into the liner cavity of the RTG. A crane operator controls the rate of descent of a GPHS stack into its cavity so that the rate of change of temperature sensors embedded in the RTG does not exceed a pre-determined limit. Thermal shock to the thermoelectric couples could occur if the pace of the rise of temperatures exceeds the temperature change limit. This process can take several hours.

The fueling end interface plate is secure on the top of the stack and a resistance check is performed to verify the heat source stack has a grounding path that prevents charge build up. Preload is applied to the GPHS stack in the cavity by a layer of insulation on top of the fueling end interface plate under the RTG fueling end cap. Figure 7.10 shows the preload being applied to the RTG.

The interface plates and preload ensure that the heat source stack does not shift during acceptance testing and mission events. A special tool that contains a load cell and digital read-out is used to apply the specified preload to the stack. Other RTGs use different fuel restraint methods and require different preload values. Threaded fasteners along the perimeter of the end cap of the RTG are torqued into place to secure the preload.

Figure 7.10 RTG assembly fixture applying preload to the end of an RTG fueling end cap. Credit: US Department of Energy.

The RTG, connected to a gas management system in the IAAC, is backfilled by exchanging argon with helium inside the generator. The exchange port is welded shut and a helium leak check is completed. These welds need to be completed in a timely manner to minimize the amount of helium lost from the generator. Helium gas provides optimal heat transfer from the hot GPHSs to their encapsulating cavity. A helium leak verifies the fueling operations are complete. The generator must leak at a rate to vent the generated helium but prevent as much air ingress as possible during ground operations.

The fueled RTG is removed from the IAAC, and the height of the RTG is measured and a map of gamma rays radiating from the RTG is created. Electrical checks of the power output at 28 VDC and instrumentation isolation occur in ambient air. The "ambient air test" is a metric that provides an overall health check of the power output from the RTG. Power circuit and instrumentation isolation tests capture the electrical resistances between an RTG's housing and thermoelectric circuit and between platinum resistance thermometers and a thermoelectric circuit. and measure its power output. These tests also provide temperature readings from the RTG's temperature sensors.

The fueled generator now undergoes a series of performance and environmental acceptance tests to accept the generator for a flight mission. Each test plays a part in INL's larger objective to verify a generator will survive a launch and the mission, have predictable effects on the spacecraft and mission, and deliver power under "mission-like" environmental conditions. The tests, their results, and the data gathered are controlled through testing requirements and plans that are developed for an RTG and a specific mission. All the processes use rigorous operations and quality control processes.

The first tests performed on a fueled RTG are vibration tests. The primary technical aim of these tests is to prove that an RTG can withstand and survive vibration levels that it will experience during launch. INL's vibration test station comprises a 9-1/2 ton Ling Dynamic System (LDS) 980 Shaker with an auxiliary cooling unit, power supply, power amplifier, impedance matching transformer, and a Kimball slip table. Aside from some system and component upgrades, the major pieces of the vibration equipment are of a 1980s vintage and were moved from the DOE's Mound Laboratories in Ohio, to the Department of Energy's (DOE's) Idaho National Laboratory in Idaho Falls, Idaho, in the early in the turn of the century. In Figure 7.11, an INL vibration operator monitors the vibration testing operations and data monitoring and collection during an RTG vibration test.

The equipment can perform vibration tests of all three axes of an RTG and can achieve acceleration levels of 100 g, a maximum force of 30,000 pounds, and a maximum displacement of 1 inch. A horizontal 2-inch thick magnesium slip plate floats on a thin, pressurized oil membrane that forms a near frictionless hydraulic bearing between the slip plate and the granite block that is the base of the shaker.

Figure 7.11 INL vibration technician monitoring RTG test date displayed on the vibration software monitoring equipment. Credit: US Department of Energy.

The entire vibration system uses pressurized pillows that allow leveling of the system and isolates the system from the building's floor and surrounding structure. An RTG is mounted onto the slip plate using a mounting fixture that bolts to the RTG and to the slip plate. The fixture incorporates triaxial piezoelectric force transducers and piezoelectric accelerometers to collect and report forces and accelerations during the vibration tests. [4]

All three axes experience vibration and sine burst tests with defined frequencies and accelerations. Test instrumentation, software, and a table control system are used to provide inputs to the shaker and capture response test data. Ground support equipment and an auxiliary high-speed oscilloscope connected to the RTG monitors power, voltage, and current. An RTG mounted to the vibration table in the vertical orientation is shown in Figure 7.12.

Results are required to show specified random vibration levels and quasi-static (i.e., sine burst) levels were achieved on all three axes tested. Test data evaluated in three main ways: did power output remain unchanged throughout the test, were any shifts of the fundamental frequency less than 5%, and last, were any irregularities in voltage or current seen during any test that could show there is a potential for thermoelectric module(s) shorting during launch? [4] Final verification occurs when the thermoelectric power circuit and internal instrumentation are shown to be isolated and that the power output of the generator matches the levels measured before the vibration tests.

Figure 7.12 An MMRTG mounted to its vibration test fixture in the vertical test orientation and configured with force transducers and accelerometers. Credit: US Department of Energy.

An RTG is one of the components taken into consideration in calculating the mass properties of a spacecraft. Accurate knowledge of the mass properties of an entire spacecraft and its components is important for modeling and understanding spaceflight dynamics. A related reason is to understand how the mass properties of spacecraft components, such as an RTG, contribute to the mass properties of the entire spacecraft. The configuration of the RTG and its mount could be a considerable distance from the center-of-gravity of a spacecraft and therefore significantly contribute to the mass properties of a spacecraft. [5] Mass properties measurements include several distinct measurements: mass, location of the center-of-gravity, and moments of inertia.

A test of an RTG at INL can determine mass and center of gravity (CG). The mass of an RTG must not exceed a value provided by mission integrators and specified to not cause anomalous spaceflight dynamics.

A calibrated scale is used to measure a fueled RTG's weight to a very high accuracy and precision. An overhead crane is used to move an RTG onto the scale. Some pieces of non-flight hardware attached to an RTG during the measurement increase the weight and their weight must be subtracted from the total weight. Figure 7.13 shows the weighing operation of the RTG.

Figure 7.13 An RTG being weighed. Note the gold and red rigging for lifting the RTG. Credit: US Department of Energy.

One challenge to measuring mass properties is controlling disturbances. At INL, personnel not needed for the weight measurements are excluded from the room during the measurements. Ventilation of the room could affect the measurements and is therefore disabled. A weighed RTG is transferred to the mass properties testing station next.

The equipment used to find the center of gravity (CG) of an RTG is a KSR1500 manufactured by Space Electronics. The instrument locates a CG to within 0.025 mm (0.001 inches). The mounting fixture for the KSR1500 has floating mounting pads, each with a precision dial used to ensure that the center of the mounting hole pattern is over the center of the instrument table.

The heat given off by a fueled RTG causes thermal expansion of the fixture, and the expansion is compensated for with the floating pads. [4] An RTG can then be rotated into a variety of horizontal orientations, vertically rotated through a full 360 degrees and rotated at various angles to locate the CG. Figure 7.14 shows the RTG configured in a horizontal position during the center-of-gravity measurements.

Once all the measurements are complete, the RTG is removed from the mass properties test station and the tests are repeated with a "bare", or empty, fixture. The "bare" measurements are used in the analysis to locate the CG of an RTG, and if needed by a mission, the moments of inertia. [4,5]

The electrical current flowing through an RTG produces a magnetic field. Characterizing and measuring the magnetic field is important in understanding

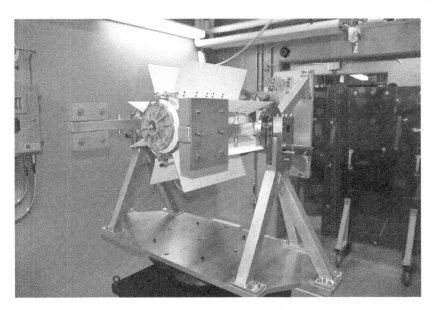

Figure 7.14 An RTG mounted to the custom aluminum fixture for mass properties measurements. Credit: US Department of Energy.

the affects the field has on a spacecraft's electronic systems. Any requirements to conduct tests on an RTG's magnetic field are specified by a mission. Measurements of the magnetic fields of RTGs have been done for many, but not all, RTGs in the last several decades.

The primary aim of measuring the magnetic field is to show that the RTG meets the requirements, but more critically, that the RTG field should not imbalance or cloud measurements made by a mission's instruments. An RTG's field is sensed and measured when a specified DC current is flowing through the RTG. Four tri-axial, digital fluxgate magnetometers are used to gauge the field of an RTG at points and distances around, beneath, and above a generator. The measurements require a very controlled environment to eliminate or minimize background noise that could give false readings during the testing. Access to the facility, where the test is being conducted, is restricted, and auxiliary facility systems are secured, and movement of personnel in the vicinity is minimized while data is being collected. Ground support equipment connected to the generator is shielded to minimize any coupling to the magnetic field measurements. A sequence of measurements is repeated twenty-five times to form a full set of spherical measurements surrounding an RTG. The INL magnetic field testing station is shown in Figure 7.15 with the RTG mounted to the test stand.

Figure 7.15 An RTG positioned for magnetics measurements. Credit: US Department of Energy.

A test in a vacuum is required to further validate and predict how an RTG will perform in a controlled, "space- like" environment. The test and equipment must be capable of sustaining a very low atmospheric pressure, near vacuum, and temperature, also known as sink temperature, to verify power values.

Two thermal vacuum atmosphere chambers (TVACs) are used at INL. Each chamber is 6 feet in diameter and 9 feet long and capable of achieving vacuum levels of approximately 1.3e-5 Pa (1e-7 Torr) in 4 hours. The chambers use cryogenic and turbo-molecular pumps. The cryopumps provide pumping with the added benefit of pump isolation from the chamber volume, thus eliminating any chance of contaminating any hardware in the chamber with oil. The turbo-molecular pumps support the cryopumps by evacuating gas species that are difficult to condense with the cryopump. Figure 7.16 shows the INL TVAC chamber used for testing RTGs.

The thermal control system of one of the TVAC chambers uses a shroud that lines the inside of the chamber with the capability for cooling and heating to control heat sink temperatures from -185 to 110 degrees Celsius (-301 F to 230 F). [6] Gaseous nitrogen flows through the shroud. This gas can be either heated with electrical heaters or chilled using the latent heat of vaporization of liquid nitrogen to control the gas within a precise temperature range.

The objectives of TVAC testing for an RTG is to verify operation in mission atmosphere conditions and validate numerical models for the power output of the

Figure 7.16 Thermal Vacuum Atmosphere Chamber to test RTGs. Credit: US Department of Energy.

flight unit. Thermal-vacuum tests verify three key requirements: first that an RTG's power output meets the mission needs while operating in a vacuum; second, that the temperatures at the base or roots of the RTG's fins remain within design limits; and third, it shows the RTG can operate over its load voltage range.

Prior to performing any TVAC testing, thermocouples are installed on the outside of the housing of an RTG and a test at ambient air temperature is performed before an RTG is loaded into a chamber. This establishes an RTG's performance at a variety of load voltages, and the data gathered in this test establishes an RTG's performance for later comparison with its performance after a TVAC test. An RTG is supported within the TVAC suspended from an overhead trolley and rail system. An RTG is shown in Figure 7.17, supported inside the INL TVAC in preparation of conducting the TVAC testing.

Feedthroughs on the TVAC wall allow monitoring of the chamber and its instrumentation. The feed-throughs also allow ground support equipment to be mated with an RTG. Controlled evacuation of a chamber ensures the health of an RTG. A controlled pump down rate is used to prevent any damaging temperature changes. Critical temperatures such as hot junction temperatures of the thermoelectric couples, fin root temperatures, and the rate of change in hot-junction temperatures are monitored during pump down. All three measurement streams are used to prevent damage.

Figure 7.17 An MMRTG suspended in a Thermal Vacuum Atmosphere Chamber. Credit: US Department of Energy.

Each RTG is subjected to a parametric test which characterizes performance at various voltages over a defined range once the chamber is at the desired pressure and temperature. The vacuum chamber pressure obtained for testing the RTG was 1.3e-5 Pa (1e-7 Torr), as an example. Each transition to a different load voltage is followed by a wait period to allow the RTG to reach thermal equilibrium with the chamber's sink temperature. Thermal equilibrium minimizes errors in power measurements induced by temperature changes. Data collected on electrical power, open circuit voltage, and temperatures at each load voltage are recorded. Compiled data are compared against previous data, performance predictions, and thermal models to verify requirements and an RTG's performance. Vacuum is broken following completion of parametric testing, and the chamber is filled with air until the chamber reaches atmospheric pressure and temperature. Subsequent ambient air and electrical isolation tests provide measurements for comparison with data collected before the thermal vacuum test.

Sidebar: The fueling and assembly of the first MMRTG at INL finished in late 2008, and the first environmental acceptance tests had begun. This first of a kind MMRTG, designated as Flight Unit 1, was going to Mars aboard the Curiosity Rover to power NASA's Mars Science Laboratory mission slated to launch in the fall of 2009. In early December 2008, NASA announced a delay in the launch of the MSL mission and Curiosity for two years because of budgetary and technical issues with several of the mission's flight components and equipment. The two-year delay presented a challenge. Environmental testing was underway. Should the remainder of the tests be postponed for two years or should testing be carried on and finished? Testing continued, and the MMRTG was put into long-term storage after the tests to await the new launch date in 2011. Requirements for storage of the MMRTG had to be developed with the goal of establishing storage requirements and scenarios to ensure its safety. Concerns with storing the fueled generator for an extended period revolved around the potential degradation of the thermoelectric couples in the MMRTG and relaxation of insulation in the generator that would diminish the preload on the GPHS stack. Storage requirements were developed to minimize power degradation from prolonged storage. The MMRTG could either be placed on short or on 28 VDC load. While in one of these electrical states and depending on the duration of storage, monitoring of the internal PRT temperatures and collecting the data was required. To maintain the fin root temperature of the MMRTG, auxiliary fan cooling with high-efficiency- air-particulate (HEPA) filters were strategically placed around the generator and each fan's speed controlled to maintain the desired fin root temperature. which lowered hot-junction temperatures in the RTG, thus reducing the thermoelectric couple degradation. Uninterruptible power supplies connected to the fans guarded against a facility power loss and, therefore, loss of cooling. Any alarms issued by ground support equipment were logged during mandatory daily checks. Figure 7.18 shows the RTG in the storage configuration, with active cooling being applied to the RTG.

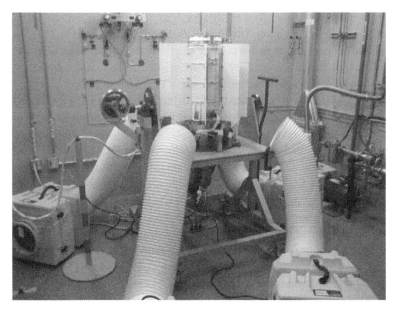

Figure 7.18 MMRTG Flight Unit 1 long-term storage preparation at INL. Credit: US Department of Energy.

The MMRTG Flight Unit 2 was fueled, assembled, tested, and stored for the Mars 2020 Perseverance Rover at INL under the identical conditions mentioned above. The Mars 2020 mission did not experience a launch delay. The MMRTG was ready for shipment to the launch site several months ahead of the planned delivery, thus the need for implementing the storage requirements.

7.3 RTG Delivery, Spacecraft Checkout, and RTG Integration for Flight

Transportation of an RTG is rigorously controlled, scrutinized, and requires specialized equipment, reams of documentation, a multitude of personnel and expertise, and many reviews and approvals by multiple governmental agencies. In the end, that all takes coordinated planning and budgeting years before a launch. INL has been responsible for packaging, transporting, and delivery of RTGs to the mission launch site since the Mound facility was closed and its operations moved to Idaho.

The first critical aspect of transporting a nuclear powered generator is to load it into a containment system that is secure and provides protection in accidents. The containment system must be analyzed, tested, and certified against credible transportation accidents to protect and prevent nuclear material from being released. The containment system certified for transporting RTGs is known as a 9904 Type B shipping cask. It was designed to transport RTGs, and the Westinghouse Hanford Company certified the shipping cask for the DOE in 1996. The shipping cask has an active cooling loop built into the exterior containment wall that connects to an auxiliary chiller system. The cask can maintain an RTG at a safe and desirable temperature while in transit. The cask is a double containment vessel per the requirements under the Code of Federal Regulations for shipping plutonium. The inner containment vessel (ICV) is a stainless steel vessel that can be sealed and then filled with an inert gas. Inside of this containment vessel are two custom made mounting fixtures that securely anchor the RTG to the shipping container. One of the RTG mounting fixtures contains passive shock indication devices that, when insulted, release ball bearings that are supported with differing springs designed to indicate the severity of the force of insult. The shock indicators are inspected after opening the shipping container to determine if such an insult has occurred. The outer containment vessel (OCV) is made of stainless steel and rests within an impact barrier that protects the OCV should a transportation accident occur. The shipping cask contains a feedthrough to allow an RTG to be monitored during transit. The cask is also mounted to a shock absorbing suspension system.

A second critical aspect of transporting a nuclear powered generator is the very means of transporting the 9904 shipping cask across thousands of miles from Idaho to a launch site. INL is the custodian of two custom semi-tractor trailers. Each is known as a Radioisotope Thermoelectric Generator Transportation System (RTGTS). The trailers meet all regulatory requirements for transporting nuclear materials within the US. Each RTGTS can transport one 9904 shipping cask in compliance with nuclear safety analysis and requirements, while the cask is loaded with a nuclear device. Each trailer is equipped with redundant auxiliary cooling systems for the 9904 shipping cask, and each trailer has two diesel generators that provide power to various electrical systems in the trailer when it is not connected to shore power. One generator can provide power for all the electrical systems in a trailer. The second generator is programmed to power up and supply power if the first generator shuts down. The trailer is equipped with a monitoring and data acquisition system that provides telemetry on the trailer's systems, and it monitors the temperature of an RTG inside a 9904 cask. A heating and air conditioning system is built into the trailer to maintain the comfort of the operators when working inside the trailer. There are two secure doors on either end of the trailer and two sets of side doors for loading and unloading shipping casks. Built into the roof of the trailer are hatches for emergency access to the casks should there ever be a situation where the side doors become obstructed or damaged. [7]

An RTG is packaged in a 9904 shipping cask at INL's Space and Security Power Systems Facility (SSPSF). An overhead crane within the high bay area of the SSPSF is used to lift an RTG onto its mount on the base of the cask's ICV. The RTG is shorted and remains shorted during transportation. A portable data acquisition system collects temperature data from the MMRTG. The ICV cover is secured, sealed, and the ICV backfilled with helium gas to maximize the heat transfer from the generator to the ICV walls. The ICV seal is leak checked. Operators lower an OCV over the ICV and bolt it down, and the auxiliary chiller system is connected to the OCV to cool the cask while it awaits loading into the RTGTS. The OCV is purged using a backfill of helium gas. That is then leak checked. The OCV is hoisted onto an impact limiter that provides both a protective barrier against impacts and a thermal barrier against a fire due to an accident during transportation. The cask is hoisted and mounted on a shock absorbing suspension system. Clearances between the door frame of the RTGTS and the sides and top of this fully integrated cask are tight and leave little margin for operators. The lift is via a forklift. Once inside and on the floor of the RTGTS, the cask is bolted to the floor and cooling is delivered to the OCV if needed. The RTGTS's data acquisition system is connected to the cask to monitor and record temperatures of the RTG.

The MMRTG, once secured inside the RTGTS, is ready for transport to the launch site. Transportation requires that a support team of INL specialists travel with an RTGTS to provide expertise should anything occur in transit. The RTGTS's data acquisition system communicates out of tolerance conditions to the RTGTS chase team and driver by sounding an alarm. Training for specialized responses is necessary should there be any off-nominal conditions with an RTG or RTGTS.

Nuclear safety and security within DOE complexes is fundamental to conducting nuclear operations. Long ago, DOE established stringent rules, regulations, and directives as a necessity in its role and responsibilities in securing nuclear material and protecting human health and the environment from the dangers that nuclear material can pose. In 1980, DOE issued Order 5480.1 to replace the interim management directive that was put in place when the Department was created in 1977. The original order established environmental protection, safety, and health protection for DOE operations and all its contractors. Since 1980, the Order has undergone several revisions and its purpose has been expanded to impose requirements that the DOE's contractors develop safety analyses that underpin evaluations of the adequacy of the safety bases of their facilities. A Nuclear Safety Analysis Report (SAR) documents the results of analyses of the safety bases for a facility. Implementing this order is specified within DOE-STD-1027, which defines and establishes guidance for the preparation and review of hazard categorization and accident analyses techniques. The standard focuses on (1) identifying nuclear facilities required to have SARs, (2) the SAR implementation plan and schedule, (3) the methodology for hazard categorization to apply to nuclear facilities, and (4) the accident analysis techniques appropriate for the graded approach defined in the DOE order. [8]

INL assesses, categorizes, and documents nuclear safety envelopes and technical safety requirements for hazardous nuclear operations performed away from INL and involving an RTG. Off-site operations, for example, are the operations that INL performs with an RTG at the Kennedy Space Center (KSC) and Cape Canaveral Air Force Station (CCAFS) to support integration. A team of analysts conducts a thorough review and analysis of all potential and credible accident scenarios and then establishes requirements for ensuring the potential accident scenarios are mitigated. These analyses are captured in a Documented Safety Analysis (DSA) that is the basis for establishing nuclear facilities at the Kennedy Space Center and Cape Canaveral Air Force Station. A DSA deems the risks of the nuclear operations (both normal and off-nominal) as manageable, controlled, and safe to perform.

The Radioisotope Thermoelectric Generator Facility (RTGF) is in the KSC Industrial Area. Built in 1964 as an ordinance storage facility for NASA, it now serves as the facility that INL operates for the receipt, flight component integration, electrical testing, and interim storage of RTGs. The facility has a fence for security and an access control point that is activated once an RTG arrives at the facility.

INL supports all integration activities conducted by a mission that involves a radioisotopic generator. RTG operations begin when the RTGTS, carrying the 9904 shipping cask with an RTG, arrives at KSC's RTGF. KSC's Launch Services Program (LSP) coordinates and brings together all the mission stakeholders and support organizations to launch a NASA mission and has been doing so since 1998. INL coordinates and interfaces with various organizations that LSP subcontracts with to conduct all RTG related operations. This includes KSC support organizations such as heavy equipment operators, health physics oversight, safety organizations, security and equipment transportation groups, and facility operators and maintenance personnel. INL works with NASA's LSP to perform the offloading of an RTG from an RTGTS.

A forklift is used to unload casks from an RTGTS operated by KSC's Test and Operations Support Contract (TOSC) operators. A forklift operator moves the shipping cask into the RTGF and its one large primary room through a roll-up door. This room contains a 5-ton dual hoist, double girder bridge crane for all of INL's hoisting operations in the RTGF. In Figure 7.19, below, INL operators are performing the 9904 Cask disassembly.

Figure 7.19 INL nuclear operators disassemble a 9904 shipping cask. The white MMRTG Flight Unit for the Mars 2020 mission sits on the base of the cask. Credit: US Department of Energy.

With the cask disassembled, the RTG was lifted from the base of the cask and moved to INL's Converter Shipping Container Base Assembly (CSCBA). The CSCBA serves as the primary means of platform for staging and storing an RTG at various facilities and serves as the support stand and enclosure for RTG transportation on KSC and CCAFS. GSE is connected to an RTG by technicians and they perform electrical checks, isolation resistance measurements, and run a 28 volt load, ambient air test to establish that an RTG did not experience any insults during transport that may have affected its functionality. Those power measurements verify the health of an RTG.

The Jet Propulsion Laboratory (JPL) designed and operated the Mars Science Laboratory (MSL) and Mars 2020 missions for NASA and handled payload and flight system integration of an MMRTG onto each rover. INL worked with the missions's Assembly, Test, and Launch Operations (ATLO) teams and coordinated the delivery and supported the integration of the MMRTGs. The first integration installed flight hardware, a flight adapter, and flight cable, to the MMRTG. The ATLO teams used a custom lift and turnover fixture (LTOF) to pick up and orient the RTGs. The LTOF allowed operators to rotate both RTGs into a horizontal position to secure a flight adapter to the mounting end of each MMRTG–the flight adapters that were the mechanical interfaces between the MMRTGs and the rovers. Installation of an electrical cable was the last mate to each MMRTG's electrical connectors. Any issues, pin damage, or defects are caught during this final mating. As with operations conducted at INL during testing, the flight cable installation requires an RTG to be placed in an open circuit condition. The open circuit condition breaks the electrical circuit between the RTG and INL's portable monitoring electrical equipment and removes the load on the output of an RTG. This prevents electrical arcing across the connector pins of an RTG connector and allows installation of a flight cable. All of these flight hardware installations occurred inside the RTGF, and quality assurance experts from INL and JPL were involved in the operations. Therefore, this operation is a timed and well-coordinated operation to minimize the duration of the open circuit condition and thus limiting a rise in hot-junction temperatures of the thermoelectric couples in an RTG.

A key risk reduction during flight integration activities at KSC is an operation termed the "hot-fit" check. This check is the first time that a nuclear fueled flight RTG gets connected to a flight rover or spacecraft. This check is months before launch, so any issues that arise can be corrected prior to launch. Recent "hot" fit checks took place in KSC's Payload Hazardous Servicing Facility (PHSF). Built in 1987, the PHSF is used by many of NASA's missions for finishing spacecraft servicing and integration. It is in a secured area surrounded by a fence and controlled access doors. The building is a voluminous structure containing two clean rooms with 100 foot ceilings. Those rooms are the hazardous operations service bay and the service bay airlock. The service bay meets the cleanliness requirements of a level 4, class 100,000 clean room, and the airlock meets the cleanliness requirements of a level 5, class 300,000 clean room. NASA granted the MSL and Mars 2020 spacecraft exclusive use of the facility while they were preparing to be launched. Much of the last inspection, system checkout, component assembly, propellant fueling, etc., occurs inside the PHSF. Components for the two Mars missions included a cruise stage, backshell, heat shield, and the entry, descent, and landing systems, and of course, the rover. There are a variety of hazardous elements associated with the work that is performed within PHSF, loading liquid rocket propellant and other pressurized flammable and explosive gases. Hazards within the PHSF are all analyzed as part of the documented safety analysis INL conducts for operations involving an RTG. The service bay and airlock each contain a 50-ton overhead crane for hoisting payload, spacecraft components, the MMRTG, etc.

The "hot-fit" check for the Mars 2020 mission took place approximately three months before launch. The CSCBA is used to transport the RTG from RTGF to the PHSF. The RTG is mounted to the base of the CSCBA and a protective cage is lowered over the RTG and affixed to the base of the CSCBA. The protective cage is an expanded metal mesh that protects an RTG from foreign debris and allows natural convective around an RTG and thus no auxiliary cooling is used when transporting an RTG in a CSBCA. Operators load this entire assembly into an RTGTS, as shown in Figure 7.20.

Figure 7.20 INL nuclear operators load the MMRTG Flight Unit 2 (F2), for the Mars 2020 mission, into the RTGTS in preparation for the transport to PHSF for the "hot-fit" check. Credit: US Department of Energy.

The RTGTS is driven to the PHSF a short distance down the road. The CSCBA is unloaded at the PHSF and placed inside the airlock. The airlock seals and undergoes a cleaning that exchanges the entire volume of air in the room several times until cleanliness is achieved. The doors separating the service bay from the airlock are opened. INL personnel supporting these operations don clean room gear and enter the PHSF service bay through an air blast cleaning room. The protective cage is removed and the CSCBA base with the RTG is moved into the service bay where the ATLO team connects an MMRTG to the LTOF and secures that to a cart called the RTG Integration Cart (RIC). The RIC is used to orient an MMRTG at the angle required for integration at the launch pad. The RIC has multiple degrees of freedom and allows for various translations of an RTG to enable precise alignment with the rover. The "hot-fit check" of the MMRTG to the Mars 2020 Rover is shown in Figure 7.21. INL's electrical team sets up the MMRTG electrical Ground Support Equipment (GSE) in preparation for RTG integration to the rover. This work is completed in tandem with the ATLO team completing the transfer of the MMRTG to the RIC. Then the MMRTG is positioned and aligned to the rover.

With the MMRTG in position to mate to the rover, the INL GSE is connected to the flight cable, and the RTG is placed on a 28 volt load by the GSE. Once the load is applied, the Heat Rejection System (HRS) team mates an auxiliary cooling system to the secondary cooling loop on the MMRTG. The secondary cooling loop carries refrigerant to drive down the fin-root temperature of the MMRTG and the

Figure 7.21 MMRTG Flight Unit 2 (F2) for Mars 2020 mission "hot-fit" checks at KSC's PHSF. Credit: US Department of Energy.

primary cooling tube. This technique is used so no overheating of spacecraft components occurs within the aeroshell during final integration and that the RTG and the spacecraft cooling systems can be mated. The refrigerant introduced into the spacecraft tubing does not boil when it flows into the warm RTG tube because the secondary tube cooled the primary tube. [9] The "hot-fit" check gives the HRS team the opportunity to test out the secondary cooling system, perform system leak checks, and determine cooling parameters necessary to achieve the desired fin root temperature on an RTG. Following a cooling system leak check, the HRS team injects the fluid known as Galden HT-170 into the secondary cooling loop on the MMRTG. While the fin-root temperature decreases with the working fluid system removing heat, the INL team monitors the portable monitoring package (PMP) for the critical parameter of the thermoelectric hot-junction change rate and to ensure the change rate is not exceeded. The HRS team continued to remove heat until they achieved the desired fin root temperature. The mechanical fastening and electrical mating of the RTG to the rover is conducted by the JPL ATLO team when temperatures have stabilized at desired values. Electrical power on and off measurements and numerous voltage measurements are made by JPL's electrical team. INL is responsible for setting specific electrical loads, using the PMP, to achieve the required current to support JPL's electrical measurements. The goal of electrical integration is to transition electronic load control from INL's GSE to the rover in a very systematic and controlled manner. INL's role in these operations is complete once power transitions to the rover until the process is reversed during demating operations. Demating took place the following day for the MSL Curiosity and Mars 2020 Perseverance Rovers, making each "hot-fit" check a 48-hour test. Overnight, the ATLO team conducted several power functional checks of the rovers, checks of RTG effects on instruments and subsystems, and telemetry and command checks of the rovers. INL provided only nuclear safety and nuclear facility management overnight, that is, no external INL monitoring of an RTG is needed once the generator is mated with a spacecraft. The ATLO team performs the monitoring.

Following the "hot-fit" check, the RTGs were transported back to RTGF to await final launch integration activities. More of the same electrical checks, isolation resistance measurements, and ambient air tests are performed as a precaution. This was an assurance that nothing abnormal had occurred to the MMRTGs following the "hot-fit" checks and the transporting back to the RTGF. The RTGs remain in storage at the RTGF until final launch integration activities commence. The RTGF provides a secure staging and storing area, while the INL team awaits the final launch integration. Continuous security support comes from KSC's LSP organization. INL provides a two-person support team at KSC for nuclear safety while the MMRTG is stored at RTGF. Daily checks of any RTG and the RTGF are performed with the support of KSC's health physics specialists that provide radiation protection and monitoring.

The final integration of the MMRTG was kicked off for the Mars 2020 mission a few weeks prior to the launch window opening. A final electrical check and 28 volt ambient air test is conducted. On the day of integration, INL loads the RTG into the RTGTS and transports it to the launch site on CCAFS. The United Launch Alliance (ULA) operates the Vertical Integration Facility (VIF) at LC-41 and handles all operations related to ATLAS V launch vehicles. The VIF is a 282-foot tall structure measuring about 75-foot by 75-foot. There are several fixed and moveable work platforms at various heights from the ground up to Level 7. A 65-ton bridge crane sits atop the upper structure of the VIF and hoists rocket motors, fairings, payloads, RTGs and more. Weeks prior to the date of the MMRTG's integration with the Mars 2020 rover at the VIF, ULA encapsulated the spacecraft within the payload fairing and transported the fairing to the VIF for installation onto an ATLAS V 541 rocket. The integration of the MMRTG with the rover was the last essential integration for Perseverance.

The MMRTG was unloaded from the RTGTS and placed within the operating envelope of the VIF crane, see Figure 7.22. The VIF crane is operated a small team

Figure 7.22 MMRTG Flight Unit 2 (F2) for Mars 2020 mission being hoisted up to the VIF Level 7 for final integration operations. Credit: US Department of Energy.

of skilled and qualified operators and crane directors. INL personnel can provide hoisting direction to the VIF crane director to aid in lifting an RTG. The 65-ton crane hook latches to the RTG's protective cage.

Prior to the lift, INL personnel assessed the VIF crane to determine the adequacy and implementation of the crane operating, maintenance, and inspection procedures. A safety analysis report commitment (SARC) was written about the 65-ton VIF crane and it analyzed the potential for the crane to drop the RTG. The report concluded the RTG could be lifted safely. The SARC findings guided INL's operating procedures and authorization to proceed with the lift of the RTG. The MMRTG, mounted to the CSCBA base with the protective cage fastened to the CSCBA and surrounding the MMRTG, is hoisted approximately 180 feet into the air and set down on one of the outer mobile platforms of work platform 7. Once there, INL removed the protective cage. INL's electrical team and the GSE staged on VIF work platform 6 provided electrical mating of the MMRTG to the GSE. More electrical tests ensued. The GSE provided continuous monitoring of the RTG during its mechanical and electrical integration to the rover. The MMRTG was then turned over to the ATLO team for the last time. The same actions taken in the "hot-fit" check are repeated for integration with the rover sitting atop the Launch Vehicle (LV). The integration operations repeated with INL monitoring the rate of change of the thermoelectric couples' hot-junction temperature as refrigerant is introduced into the MMRTG secondary cooling loop and then, INL disengages the GSE electrical load and the electrical connection between the spacecraft RTG is made. The closing of the aeroshell doors and the payload fairing doors encapsulate the spacecraft and RTG for launch. In a matter of days the rocket will be pulled out of the VIF to stand at launch on the pad called SLC-41 and await the ignition of its liquid booster stage and solid rocket motors to launch Mars 2020 into space.

The cruise to Mars culminated seven months after liftoff and 300 million miles from Earth where Mars 2020 landed the Perseverance rover in the Jezero Crater on 18 February 2021. The MMRTG was healthy and was expected to provide a reliable source of power for an incredible mission with objectives of looking for habitable zones for life and biosignatures and collecting rock cores and soil samples for a subsequent sample return mission. Perseverance will also test the rover's ability to extract oxygen from the Martian atmosphere, a process vital to future human missions on Mars.

References

1 Idaho National Laboratory – Battelle Energy Alliance (2015). LLC Space Nuclear Power and Isotope Technologies Division, Atomic Power in Space II.

2 US Department of Energy (DOE) (2020). Certificate of Compliance for Radioactive Materials Package (Certificate No. 9516).

3 Giglio, J. (2020). Review of the General Purpose Heat Source Module Reduction and Monitoring at INL.

4 Geddes, K. and Dees, C. Vibration and Mass Properties Testing of the MMRTG Power System that will power Perseverance, the next rover to the surface of Mars.

5 Felicione, F.S. (2009). *Mass Properties Testing and Evaluation for the Multi-Mission Radioisotope Thermoelectric Generator*. Idaho National Laboratory.

6 Birch, J., Horkley, B.M., Giglio, J., and Veselka, H.D. (2018) Renovation of thermal vacuum chambers at idaho national laboratory for testing of radioisotope power systems. 30th Space Simulation Conference, Annapolis, MD.

7 Johnson, S.G. and Lively, K.L. (2010). *Evaluation of Storage for Transportation Equipment, Unfueled Converters, and Fueled Converters at the INL for the RPS Program*. INL.

8 US Department of Energy. (1992). *Hazard Categorization and Accident Analysis Techniques for Compliance with DOE Order 5480.23, Nuclear Safety Analysis Reports*. United States. https://doi.org/10.2172/7042507.

9 Bhandarai, P., Diduk, B., Birur, G., Bame, D. et al. (2012). *Mars Science Laboratory Launch Pad Thermal Control*. American Institute of Aeronautics and Astronautics. https://doi.org/10.2514/6.2011-5116

8

Lifetime Performance of Spaceborne RTGs

Christofer E. Whiting[a] and David Friedrich Woerner[b]

[a] *University of Dayton Research Institute, Dayton, Ohio*
[b] *Jet Propulsion Laboratory/California Institute of Technology, Pasadena, California*

8.1 Introduction

The US has used RTGs to provide reliable, uninterrupted thermal and electrical power for spacecraft since the 1960s. A detailed understanding of how power output from an RTG degrades with time is critical to planning a long space mission. For example, a science mission to an outer planet of the solar system may have a nine-year cruise followed by an intense 3-month approach and flyby. Overestimating the electrical power available will, at least, result in the mission performing fewer science experiments, and in extreme cases, could cause difficulty operating the spacecraft or mission failure. Overestimating the RTG waste heat available could cause the failure of critical spacecraft components and systems that need to be kept warm, like sensitive electronics or thruster propellant. Underestimating the power may seem preferable, but erring too far in that direction may unnecessarily limit the ambition and scope of space missions, causing scientists to reduce the instrumentation, number of experiments conducted, and/ or mission scope. Besides lost science, this could underutilize an expensive national asset. For these reasons, developing a precise understanding of the lifetime performance of each RTG has been the focus of a significant amount of time, energy, and resources within the RTG community. [1–5]

Performance degradation in an RTG ultimately comes from two sources: decreasing radiogenic heat and degradation of the thermoelectric converter. Reduction in radiogenic heat occurs because the heat producing ^{238}Pu atoms are

The Technology of Discovery: Radioisotope Thermoelectric Generators and Thermoelectric Technologies for Space Exploration, First Edition. Edited by David Friedrich Woerner.
© 2023 John Wiley & Sons, Inc. Published 2023 by John Wiley & Sons, Inc.

consumed during radioactive decay.[1] This causes a proportional reduction in heat generation of 0.787% per year. This lessens the thermal power available for conversion and reduces temperatures within a thermoelectric converter. Reduced thermoelectric converter temperatures means that there will be a decrease in the ΔT across the thermoelectric couples and a change in the integrated figure of merit, ZT.[2] This temperature change reduces RTG power output by 0.2 to 0.7% per year, for a total electrical power loss of 1.0 to 1.5% per year. The combined degradation from the reduction in fuel and associated decline in temperature are referred to as thermal inventory losses.

The second source of performance loss is degradation of the thermoelectric converter. Converter degradation is more complex than radiogenic heat loss and it depends on the thermoelectric materials, thermal/electrical interfaces, and insulation used in the converter. SiGe-based converters, for example, experience ~9% loss in performance because of converter degradation over 14 years, while PbTe/TAGS based converters experience ~25% loss over the same time. An exhaustive discussion of every degradation process that occurs in the ten[3] US RTG designs that have flown in space could be the subject of its own book. Discussion here will focus on the degradation mechanisms that are known, or suspected, to be the primary cause of converter performance losses.

Thermal inventory losses for specific RTGs are not discussed extensively in the open literature. Thermal inventory losses are easy to anticipate because the very consistent process of radioactive decay controls them. In other words, different RTGs that use the same fuel (e.g., ^{238}Pu) should exhibit similar thermal inventory losses. That means most of the differences in performance observed in different RTG designs stem from converter degradation. A history of RTG performance is, therefore, largely a history and evolution of thermoelectric converter performance.

This chapter presents and discusses the performance data available from all US RTG-powered space missions flown by the end of 2020. Some historic missions may not show the excellent reliability we expect today, yet it is a point of pride that no spacecraft has ever failed to achieve its mission because of an RTG failure or anomaly. In some instances, a spacecraft failed before it reached the RTG design life, but as later sections of this chapter will show, these RTGs were all functioning years later as seen by observing signals from instruments onboard the spacecraft. These early RTGs provided many lessons that led to significant improvements in future spacecraft RTG designs. After about 15 years of technology maturation, the

---

1 The fact that ^{238}Pu is consumed to produce heat is why the material is referred to as a "fuel." The fact that fuel is consumed to produce power is why the RTG is a generator.

2 Material property closely related to conversion efficiency. See Chapter 6.

3 SNAP-19 design iterations for Nimbus, Pioneer, and Viking are significant, and they are considered different converters for this purpose.

US began manufacturing highly reliable, extremely long-lived RTGs to power NASA missions such as Cassini (20 years), Pioneer 10 (30 years), and the Voyager probes (44 years and counting).

8.2 History of RTG Performance at a Glance

Figure 8.1[4] presents RTG power that is normalized to the power produced at the beginning-of-mission. Nearly every US spacecraft that flew an RTG is included in this figure.[5] The figure presents normalized power because flown RTGs have produced a very broad range of power. Some missions used nearly 1,000 W_e of RTG power, while others used less than 3 W_e. Normalized power makes it much easier to compare all these RTG-powered missions.

Figure 8.1 is more illustrative than informative. No format can present all this data in a way that allows nuances in performance to be seen. After 60 years, 29 US missions have flown RTGs. Some missions only lasted months (Transit 4), while others are still active after 44 years (Voyager). Some missions show a slow, gradual degradation, resulting in a curve that appears to decline gently for decades. Other missions show rapid RTG degradation that could lead to inadequate power after only a few years. Figure 8.2 shows this distinction by separating the missions based on their degradation behavior.

Despite the busy nature of Figure 8.1, it illustrates several important features of RTGs as a technology. Early RTGs that predate the SNAP-19 are presented together in Figure 8.2a, which shows the high degradation rates observed in those systems. Lessons learned from each of those designs produced improvements in subsequent designs that decreased degradation and increased reliability. The SNAP-9A, flown on the Transit 5 satellite series, is the oldest RTG design presented here. This RTG had some major weaknesses, primarily in the gas management system, which caused it to have the fastest observed degradation of all flight RTGs. The SNAP-19B, flown on Nimbus III, was the next design, and while the improved gas management system resulted in slight performance improvements, degradation was still fast enough to prevent the RTGs from lasting more than a few years. The SNAP-27 RTGs, used on several Apollo missions, were the first hermetically sealed RTGs,

---

4 The authors acknowledge that Figure 8.1 is not a practical presentation when trying to look at details in RTG data. Figures presented later in the chapter will have much more utility for readers interested in those details.

5 Performance data not included: Transit 4-A, Transit 4-B, and TRIAD. The power data for a few other missions are incomplete. This includes Transit 5BN-2 after primary function as a navigational satellite ceased, Pioneer 10 after 12/31/95, and LES 8 and 9 after 9/1/88. See subsections on these missions for more details.

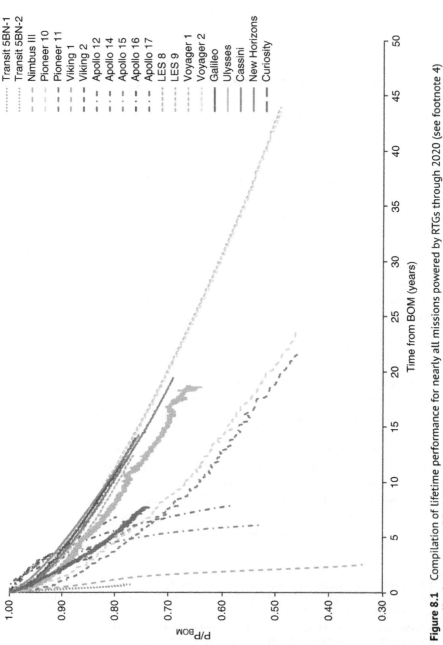

Figure 8.1 Compilation of lifetime performance for nearly all missions powered by RTGs through 2020 (see footnote 4) normalized to power at BOM. Different RTG types are represented by different types of lines: SNAP- 9A (dotted), SNAP-19 and -19B (medium dash), SNAP-27 (dash-dot), MHW-RTG (short dash), GPHS-RTG (solid), and MMRTG (long dash).

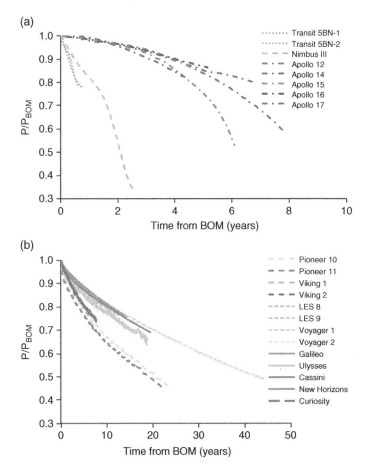

Figure 8.2 Lifetime performance for RTG missions that predate SNAP-19 (a) versus missions that flew with, or after, SNAP-19 (b), normalized to power at BOM.

and they showed low power degradation over the short-term. After a few years, however, the unexpectedly high temperatures of the lunar daytime resulted in accelerating degradation and significant power losses.

Thanks to the lessons learned from the early designs, all RTGs that followed are highly reliable power sources. SNAP-19, MHW-RTG, GPHS-RTG, and MMRTG all show graceful degradation curves, and they are presented together in Figure 8.2b. These curves are consistent and predictable. Missions that fly the same RTG design show very similar performance over time, despite any differences in spacecraft design or mission profile. RTG reliability is inherent as no RTG has ever suddenly failed, and, except for Nimbus III, no RTG has ever shown a sudden change in behavior.

RTGs are now a very long-lived power generation technology. Over 20 years of operation is common, with RTGs powering some spacecraft for over 40 years. The Voyager 1 and 2 spacecraft are the top two record holders as NASA's longest operating missions. Most RTG missions launched after Pioneer 10 were decommissioned because other spacecraft systems could no longer function (see Chapter 3 for additional details). To date, only three spacecraft were decommissioned because RTG power was too low, despite exceeding their expected mission lifetime. These spacecraft were, Pioneer 11, Ulysses, and LES 9[6]. Pioneer 11 defied expectations by operating for over 22 years, when NASA shut down the spacecraft because it could no longer power any of its sensors. Ulysses operated for 18 years and completed two extended missions before the RTG heat and power could no longer keep the thruster propellant from freezing. LES 9 operated for an astounding 44 years before the reduced RTG power and loss of primary telemetry led to the decision to decommission the satellite.

The last notable illustration provided by Figure 8.1 is that RTG performance is most closely tied to the thermoelectric converter. PbTe/TAGS materials were used in the thermoelectric couples of the SNAP-19 and MMRTG designs. RTG telemetry from all four of the spacecraft using SNAP-19s (the Pioneer probes and Viking Landers) shows similar behavior despite very different mission environments (deep space vs. Mars) and minor design modifications that enhanced the RTG for use on Mars. [6][7] Almost 40 years after Pioneer 10 and 11 left Earth, the new MMRTG, mounted on the Curiosity rover, is displaying performance trends similar to the SNAP-19s. Meanwhile, MHW-RTG and GPHS-RTG were built with SiGe thermoelectric materials. Despite some significant changes made in those RTG designs, their performance is nearly identical.[8]

What Figures 8.1 and 8.2 do not show is the conversion efficiency[9] of each RTG design. The first RTG prototype designed for space was a SNAP-3. This system showed a BOL conversion efficiency of 5.75%. Meanwhile, the MHW-RTG and GPHS-RTG designs are tied for the most efficient RTGs ever manufactured at 6.6% efficiency. This was a performance improvement of less than 15%. Over 60 years, design improvements have significantly advanced the reliability, longevity, and specific power of the technology, but conversion efficiency has not significantly

---

6 Some would include Pioneer 10 in this short list. We have chosen to not include it because continuous operation of this mission was terminated 6 years before the spacecraft stopped transmitting. See Section 8.3.6 for more details.

7 The most significant design modification for Viking was a new hermetically sealed gas management system that would reduce converter degradation during transit, optimize power upon landing on Mars, and prevent the Martian atmosphere from leaking into the thermoelectric converter.

8 MHW-RTG missions – LES 8, LES 9, Voyager 1, and Voyage 2. GPHS-RTG missions – Galileo, Ulysses, Cassini, and New Horizons.

9 Fraction of heat energy that is converted into electrical energy.

improved. The DOE and NASA have funded a few efforts to improve conversion efficiency, but none achieved success. Today, improved conversion efficiency remains a desirable, yet elusive, goal. Some early TRL technologies being developed today show promise (see Chapter 10) and may overcome the 7% barrier in efficiency in the next 10-20 years.

8.3 RTG Performance by Generator Type

It is very difficult to get detailed information on individual RTG performance from Figure 8.1. Some RTGs display significantly different performance and lifetime, while other RTGs appear so similar that nuances in performance are not visible. Hence, a deeper dive into performance follows.

8.3.1 SNAP-3B

SNAP-3B was the first RTG to fly in space. It was a small, 2.7 W_e supplementary power source for the Transit 4-A and Transit 4-B satellites. The Transit 4 satellites did not telemeter RTG power, and it is unclear if this was a limitation of the RTG or the spacecraft. SNAP-3B telemetry that is in the public domain includes only the load voltage, main battery voltage, and outer-shell temperature. [7] Transit 4-A suffered a telemetry transmitter failure one month after launch. [8] A high-altitude nuclear weapon test irradiated Transit 4-B nine months after launch, which caused the primary power source, a solar array, to degrade extremely rapidly. [8] Communication with both satellites was lost within the first year of flight. The lack of power data and the short mission lifetimes make it difficult to evaluate the SNAP-3B.

Evidence of RTG power output was observed long after the Transit 4 satellites stopped communicating. The RTG provided power to amplifiers for 54 and 324 MHz Doppler transmitters. According to a report published in 1978, these transmitters were "still operating" on the Transit 4-A satellite 17 years after launch. [8] Signals were received from Transit 4-B in 1971, over 9 years after launch. These signals proved the RTGs had not failed, and that some meaningful amount of power was still being produced.

8.3.2 SNAP-9A

While the SNAP-3B system was successful, the power output from RTGs had to be increased before they could become the sole source of power for a mission. The SNAP-9A design produced almost ten times more power than the SNAP-3B and was the only electrical power source on the Transit 5BN series of satellites. Figure 8.3 presents performance data from Transit 5BN-1 and Transit 5BN-2. Performance data for the SNAP-9A were taken from Ref 7.

The lifetime of the Transit 5BN-1 satellite was very short. Three months after launch, the two primary telemetry bands stopped transmitting. The auxiliary telemetry band continued to function and indicated a major short-circuit either in one of the satellite wiring harnesses or in the electronics. [7] The excess load from this short prevented the Doppler transmitters from receiving enough power to operate. Five months later, Transit 5BN-1 stopped transmitting all telemetry.

Transit 5BN-2 was the first fully operational satellite used for terrestrial navigation. After 11 months of operation, however, anomalies in the satellite memory prevented the satellite from operating as a navigational system. Despite these issues, SNAP-9A performance data was still being transmitted by the satellite. According to a 1978 report from the Johns Hopkins Applied Physics Laboratory, the satellite was still transmitting good SNAP-9A telemetry data 15 years after launch. [8] When more information was requested for the writing of this chapter, staff members at the Applied Physics Lab were gracious enough to look in their archives, but could not find any records of this data. It is possible this data was not recorded, since it occurred long after the navigational system failed. It is also unlikely there will be enough interest to spend resources hunting for a potentially non-existent data set from a long defunct RTG design. Thus, it is likely the Transit 5BN-2 power data past seven months will never be found.

Transit 5BN-3 failed to achieve orbit. Thus, despite being counted as one of the 29 RTG flights that were launched, there is no performance data for the mission.

We can make two observations using Figure 8.3. First, performance of the Transit 5BN-1 RTG shows a slight deviation at three months (0.25 years) when a satellite short occurred and placed a load on the RTG, taking away power needed for other spacecraft subsystems. Three months after the short (~0.5 years in Figure 8.3), the RTG appears to have returned to normal power production. While

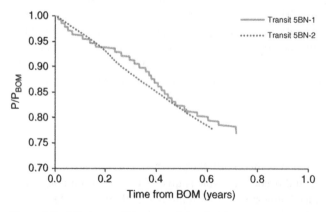

Figure 8.3 Lifetime performance of the SNAP-9A flight systems normalized to power at BOM.

there is not enough data to discuss any long-term implications of a major electrical event that knocked out one of the satellites, this similarity in short-term RTG performance showcases the resilience of RTG technology.

A second observation is the rapid decrease in performance over a short period for both RTGs. Figure 8.2 shows that SNAP-9A had the fastest observed degradation of all flight RTGs. The literature does not discuss this high degradation rate. This is most likely because the cause of degradation was not studied in depth. RTG designs were rapidly evolving due to breakthroughs in TE technology and more efficient thermal design. In fact, the AEC was already working on the next generation in space RTG technology, the SNAP-19. Attempting to discuss the cause of the SNAP-9A RTGs' degradation would be pure speculation. It is possible to say, however, that the degradation must have slowed down significantly. Otherwise, Transit 5BN-2 would not have been producing sufficient power for the Doppler transmitters in 1978, 15 years after launch. It is also important to note that the Transit 5BN missions concluded that the SNAP-9A met all objectives, including "Demonstrate satisfactory operation and potential long-life capability of the SNAP-9A power supply." [8]

8.3.3   SNAP-19B[10]

NASA's interest in RTG technology dates back to the Kennedy administration, and the famous friendship between the NASA Administrator James Webb and the AEC Chairman Glenn Seaborg. [9] They realized RTGs were the only technology that could enable exploration of deep space and the outer planets. In 1963, a few months before the first SNAP-9A launch, NASA commissioned an RTG for a proof-of-concept flight on a Nimbus weather satellite. This RTG had to produce twice the power of SNAP-9A, which necessitated technology improvements. The AEC realized it needed to show that RTGs could be the sole power source on missions to explore deep space.

---

10 The historical record inconsistently uses SNAP-19, SNAP-19B, and SNAP-19B3 when discussing the RTGs used on Nimbus III. Originally, the Nimbus RTG design was referred to as just SNAP-19. Despite all the design changes that occurred over the lifetime of the SNAP-19, Teledyne, the system contractor, never used the 'B' designation when the SNAP-19 was in production. A consecutive chain of SNAP-19 serial numbers can be traced from the Nimbus system through Viking. [6,11,16] The 'B' designation was added later by the AEC sometime in the early 1970s. [9] While it was not explicitly stated, we suspect this addition of 'B' to the Nimbus system was to highlight these major design changes. This would hopefully separate the improved SNAP-19 from the design used on Nimbus, which experienced severe gas management issues. The '3' designation identifies the system as the 3rd set of SNAP-19s built to support the Nimbus program. [9]

Doubling the power output turned into a major driving factor for the SNAP-19 design. Rather than increase the generator's size, the SNAP-19 was designed to work in pairs.[11] This allowed the SNAP-19 to scale power up more efficiently than previous systems. Increasing total power also required roughly double the amount of ^{238}Pu fuel. This resulted in a large enough quantity of radioactive material that the AEC switched from using pure ^{238}Pu metal to ^{238}PuO$_2$ microspheres. ^{238}PuO$_2$ has a lower specific thermal power (W_{th} g^{-1}), but it is a much safer fuel form. The oxide has a much higher melting point, lower reactivity, and is insoluble in water. [10] In the event of a launch accident or abort, the oxide presents a much lower hazard to people and the environment.

Doubling the fuel also caused the AEC to make a last-minute change to their safety philosophy. The Nimbus system needed enough fuel that if the RTG burned up, the amount of fuel that might be dispersed would be a significant fraction of the plutonium present in the Earth's atmosphere. With less than six months before the RTGs needed to be delivered, the AEC redesigned the heat source so that it could survive a launch accident or re-entry and be recovered intact. [9] This turned out to be a very fortuitous decision.

The Nimbus-B satellite was the first launch of an RTG on a NASA spacecraft. Two minutes after launch, Nimbus-B veered off course,[12] and the range safety officer sent a destruct signal to the rocket. [9] The new heat source fell into the Pacific Ocean, where it was recovered, intact, at the bottom of the Santa Barbara Channel. The intact recovery design was so successful that the fuel was reused on the Nimbus III mission.

Nimbus III[13] was the first successful deployment of an RTG on a NASA mission. While a solar array produced a majority of the satellite power, the two SNAP-19B RTGs provided enough additional power to ensure the satellite achieved all mission objectives. Figure 8.4 presents the performance data from Nimbus III. Performance data for the SNAP-19B is an average of the data presented in Ref 7. Most missions report total RTG performance as a sum of all onboard RTG power, not by individual RTGs. Averaging RTG performance for Nimbus III was done so that the reported SNAP-19B performance was consistent with this practice.

Early performance of the Nimbus III RTGs showed a significant improvement over the SNAP-9As. They showed a slower, more graceful loss of power, but their early degradation rate was greater than predicted. [11] Analysis of the curve for

---

11 The original SNAP-19 was designed to produce 30 W_e, which is similar to the 27 W_e for SNAP-9A. See Chapter 2 for more details.

12 Human error in setting a guidance gyro was later determined to be the cause of the launch vehicle deviating from its course. [9]

13 Nimbus III and Nimbus-B2 are the same. Prior to orbit, Nimbus used letters, and Roman numerals after reaching orbit.

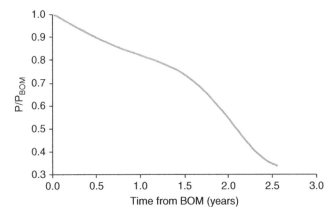

Figure 8.4 Lifetime performance of the two SNAP-19B flight systems on the Nimbus III spacecraft normalized to power at BOM.

Nimbus III shows an inflection point at 1.1 years after BOM. This is a clear sign the degradation rate had changed. Power losses on Nimbus III accelerated, and 2.1 years after BOM, the RTGs had lost over half their power.

Teledyne's analysis of Nimbus III concluded the RTG gas management system had failed. [12] Sublimation of the PbTe thermoelectric materials was a known problem at the time and the sublimation was suppressed in flight with an inert cover gas at low pressure. Loss of the cover gas would result in accelerated sublimation. Original SNAP-19B designs relied on a 103 kPa (15 psia) argon cover gas to leak out of large O-ring seals at the end caps. The argon was then replaced by helium produced during the radioactive decay of the plutonium-238 fuel. Analysis suggested that a small overpressure, 14 to 34 kPa (2 to 5 psia), would occur at equilibrium between helium production and the gas leak. [6] The analysis proved incorrect. Pressure sensors in the RTGs registered only 1.4 kPa (0.20 psia) of pressure 0.75 years after BOM. [11] After 1.1 years, the helium pressure became too low to prevent rapid sublimation of the thermoelectric couples.

Sublimation of thermoelectric materials can initiate a phenomenon called thermal runaway. Sublimation of thermoelectric material from a thermoelectric couple causes the cross section of one or both legs to shrink. This increases electrical and thermal resistance in the leg(s) suffering sublimation, which increases the temperature at the point of shrinkage. Increases in temperature escalates the sublimation rate. The resulting vicious cycle results in a runaway that can cause a system failure. The low pressures measured on Nimbus, and the inflection point in Figure 8.4, therefore, suggest that thermal runaway of the SNAP-19B RTGs began 1.1 years after launch.

It is worth noting that the SNAP-19B design life was only one year, and the generator met that requirement even though thermal runaway eventually became a problem. Despite SNAP-19B being a proof-of-concept RTG, the AEC recognized that rapid degradation after only one year of operation could jeopardize their long-term relationship with NASA. Efforts to overhaul the SNAP-19 design followed, including improvements to the thermoelectric materials, gas management system, and the heat source.

8.3.4 SNAP-27

Lunar exploration was the driving force for NASA's space exploration in the 1960s, and a major national interest for the US. It was apparent that RTGs were the only viable technology capable of providing uninterrupted power over a 354-hour-long lunar night. In 1965, NASA commissioned the SNAP 27 RTG for use on the surface of the moon. This put development of the lunar SNAP-27 in parallel with the deep space SNAP-19. The SNAP-27 RTG development marked the start of modern RTG programs in many ways. [13] Prior to SNAP-27, RTGs were custom designs developed to meet the needs of a specific mission. These early designs were for "proof-of-concept" missions, where the level of risk and tolerance for failure was much higher than allowed today. SNAP-27 was the first large budget commission that expected the technology to be reliable and available as new missions were announced. It also required the AEC and General Electric, the SNAP-27 prime contractor, to implement the rigorous development and quality assurance standards employed by most NASA missions. Failure was no longer an option.[14]

The first SNAP-27 was going to fly on the Apollo 11 mission for the first lunar landing, but there were concerns that the first crew to land on the moon was going to be overtaxed with activities. NASA postponed deployment of the first lunar RTG. Every mission after Apollo 11, however, carried a SNAP-27 to power an Apollo Lunar Surface Experiments Package (ALSEP).[15]

Figure 8.5 presents the performance for all SNAP-27 systems put into service on the moon and the data was taken from Ref 7. The data shows exceptional performance with less than a 5% loss in power over the first 2 years of lunar operation. Long-term data, however, shows that degradation sped up significantly over just a few years of operation.

---

14 General Electric originally bid the SNAP-27 program at $4.6 million. Early during development, Harry Finger (Director of the AEC Space Nuclear Systems Division) remarked that the program did not have the technology or resources needed to assure success. Finger then went to Congress and got them to increase the SNAP-27 program budget to over $10 million. [9] Finger was then quoted as saying, "Now I feel confident if you run into trouble, you will be able to fix things." [9]

15 This includes Apollo 13, which is why it is counted amongst the 29 RTG missions flown to date. The unfortunate events of Apollo 13, however, clearly prevented the RTG from being used. See Chapter 3 for more details.

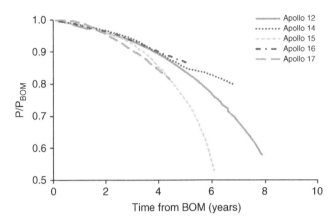

Figure 8.5 Lifetime performance of the SNAP-27 flight systems normalized to power at BOM.

This flaw emerged in the SNAP-27 because of an inadequate thermal analysis of the lunar surface. Analysis of the design established a maximum hot junction temperature of 593 °C. All nominal couple, module, and system-level testing took place at that temperature. Testing in laboratories showed rapid degradation because of sublimation of thermoelectric materials at hot junction temperatures of 660 °C [14,15], and unfortunately, the initial data from the Apollo 12 ALSEP showed that the SNAP-27 hot junction reached 617 °C during each lunar daytime. The 24 °C increase in temperature appears to be enough to initiate excessive sublimation of the thermoelectric couples. That the power losses were accelerating at the same time significant sublimation was occurring strongly suggests that the unit was experiencing thermal runaway.

Figure 8.5 shows that some of the SNAP-27 RTGs (e.g., Apollo 15 and 17) showed higher rates of degradation than the others. These differences are not discussed in the literature, and there is no clear explanation for why these missions showed faster degradation. Apollo 15 and 17 landing sites were the farthest from the lunar equator, but this does not provide a good technical explanation for faster degradation. Distance from the lunar equator should reduce daytime temperatures, which should reduce degradation. Anecdotal evidence suggests that production of the 3n/3p PbTe thermoelectric materials used in SNAP-27 was very challenging, and that some of the couple design choices can result in unreliable couples. [16] For example, SNAP-27 couples were not bonded on the hot side, and a pressure contact was maintained between the couple and the iron button with a spring. This could result in the hot end of the couple "rounding" due to sublimation, resulting in very poor contact on the hot side. Therefore, we suspect the inconsistent degradation in the SNAP-27 systems was due to sub-optimal design and production choices.

It is important to note that the SNAP-27 RTGs were a remarkable success. Each ALSEP RTG greatly exceeded its one-to-two-year design life and provided more than enough power to meet all the mission objectives. The ALSEP monitoring program was terminated in 1977 for reasons unrelated to RTG power.

8.3.5 Transit-RTG

Design of the Transit-RTG, not to be confused with the RTGs launched on the Transit missions of the 1960s, followed a unique design philosophy. The Transit-RTG was designed for operating temperatures that were low enough to stop sublimation of the thermoelectric materials (i.e., 400 °C in PbTe chemical systems). This would allow the thermoelectric materials to operate in the vacuum of space, without the presence of a protective cover gas, and simplify components that suppressed sublimation. This simpler design should have significantly decreased degradation, resulting in a more reliable and stable long-term power source.[16] The price for these improvements was a decrease in the system conversion efficiency and power output by nearly one-third (i.e., 4.2% for Transit-RTG vs. 6.2% for SNAP-19). Other RTGs had high enough temperatures to optimize power, but keep temperatures low enough to minimize sublimation and prevent system failures, like thermal runaway.

A single Transit-RTG powered the TRIAD navigational satellite, which was launched in September 1972. Unfortunately, a telemetry failure in the satellite prevented the transmission of any RTG data after 30 days. Some experiments onboard the satellite were still usable, however, so the mission continued. The only publicly available performance data on the Transit-RTG are average power values over the first 20 days (35.6 ± 0.5 W_e) and average power from 25 to 30 days (35.4 ± 0.2 W_e). [7] That is not enough data to draw any meaningful conclusions about RTG performance or lifetime. Therefore, Transit-RTG is not in Figure 8.1. Despite the telemetry failure, reports published as late as 1987 show that the magnetometers were still operating [9], which indicates significant RTG power was being produced 15 years after launch.

Designs similar to the Transit-RTG's with its lower operating temperatures and expected reduction in power losses were not pursued again. This is most likely because the lower temperature resulted in ~33% reduction in BOM power, which was too high a price to pay for "anticipated" improvements in reliability and stability. Especially when RTG reliability reached full maturity just one year later (1973) with the redesigned SNAP-19. We conclude RTG designs had become fully mature as of the SNAP-19 because no further work needed to be done to reduce

---

16 Reliable – less prone to unexpected power losses and/or failure; Stable – lower power losses over time.

the likelihood of unexpected power losses or failure.[17] Since the deployment of the redesigned SNAP-19, 27 RTGs have flown on 13 missions. None of them have ever shown unexpected power losses or failure.

In addition, the expected improvement in power stability that might have been demonstrated by a Transit-RTG did not provide enough power over the life of a mission. For example, a system like a Transit-RTG would effectively start on Figure 8.1 at $P/P_{BOM} = 0.67$. Even if it had no thermoelectric degradation (which is unlikely), decreases in power because of radioactive decay and other thermal inventory losses are ~1.2% per year.[18] Figure 8.1 shows us it would take almost 20 years before this system would surpass SNAP-19 (Pioneer), and it would probably never produce more power than MHW (Voyager) or GPHS-RTG (Cassini). Since no mission has ever planned to operate longer than 20 years, there is no reason to sacrifice that much BOM power.

There has been some work that suggests that a smaller decrease in temperature could reduce degradation in a PbTe/TAGS thermoelectric converter and produce more power at the end-of-design-life.[19] Here, a reduction in degradation is traded with a reduction of BOM power. Such a trade could result in an optimum amount of power for a mission and is discussed in greater detail in Chapter 7. This is a philosophy unique from the Transit-RTG design, which was trying to eliminate degradation in the converter.

8.3.6 SNAP-19

In 1964, Gary Flandro discovered a rare planetary alignment. This alignment occurs once every 175 years, and would allow a single spacecraft to flyby all the outer planets[20] using gravity assists. [17] This mission captured the imagination of NASA, and became known as the Planetary Grand Tour, or just Grand Tour. In order to prepare for this once in a lifetime opportunity, NASA flew a pair of probes that would be pathfinders for the Grand Tour. [17] They would be the first space-craft to travel through the asteroid belt, enter the outer solar system, and encounter Jupiter. [18] NASA knew the only way to power the Grand Tour was with RTGs, so demonstrating an RTG on these Jupiter probes became a priority. This

---

17 We note that this claim of maturity applies to PbTe/TAGS and SiGe thermoelectric converters. A new thermoelectric material may revive questions about reliability of the converter. Quality of the housing, gas management, heat rejection, and insulation, however, would still be at a high state of maturity.

18 1.2% of current power. In other words, after one year $P/P_{BOM} = (0.67 * 0.988) = 0.662$. After 2 years, $P/P_{BOM} = (0.662 * 0.988) = 0.654$. Etc.

19 Defined as 14 years of operation and 3 years of preflight storage for the MMRTG.

20 Planets beyond the asteroid belt.

sharpened NASA's focus on the SNAP-19 program, which was already underway to support Nimbus.

When Nimbus III started experiencing rapid power loss after only 13 months, the Pioneer Project started to second guess the use of RTGs. Project Manager Charles Hall became very concerned. He even looked at replacing the RTGs with solar cells [9], but RTG technology had several advantages and champions. Solar cell and array technology was much less efficient back then, and would have suffered challenges with solar intensity and low temperatures at Jupiter. NASA headquarters also favored use of RTG technology,[21] and the AEC had already begun a proactive and aggressive campaign to improve the SNAP-19. The redesigned SNAP-19 was selected for use on Pioneer 10 and 11 because of these improvements, beginning a new era for RTGs. An era where RTGs never failed to prove that they were reliable, very stable, and exceptionally long-lived.

Despite the issues observed on Nimbus III, the Viking project remained very excited about the prospect of using RTGs. Viking project management was keen to deploy RTGs on Mars. The project believed RTGs to be the only reliable solution to providing thermal and electrical power through Martian nights, winds, dust storms, and extreme cold temperatures. [9] Rapid power losses on Nimbus III did not seem to bother mission staff much. In the words of Jerry Soffen, Viking Project Scientist, "But actually, we always wanted RTGs, and we put a lot of effort into keeping the AEC in line to provide them." [9] Rather than looking into alternative power sources, like solar, Viking began working on a battery. Landing a battery provided many advantages, and one of them was to make sure the mission needs were met should the RTGs lose too much power. [6][22] Besides the improvements made for the Pioneer RTGs, Viking required several additional RTG modifications for the unique environment of Mars. [6,7,9,12] This included a gas management system that would reduce power losses during the 10-month cruise to Mars and prevent the Martian atmosphere from leaking into the RTG.

These additional modifications for Viking are the reason some literature sources discuss the Pioneer/SNAP-19 and the Viking/SNAP-19 systems separately (e.g., Ref 7 and 9) and consider them to be different systems. Because of the similarities in performance for Viking and Pioneer (discussed below), we consider the Pioneer/ SNAP-19 and the Viking/SNAP-19 to be part of the same "family" of generators.

---

21 Though it is not expressly stated, NASA headquarters most likely favored the use of RTGs because a study that used solar cells for Pioneer resulted in a "... design [that] was pretty limited." [9] They also likely favored RTGs due to the significant effort and government investment that went into redesigning the Nimbus SNAP-19B. See Section 8.3.3 in this chapter for more details.

22 The spacecraft could draw extra power from the battery when the scientific instruments were operating. The instruments could then be shut off for a short period of time and the RTG could recharge the battery.

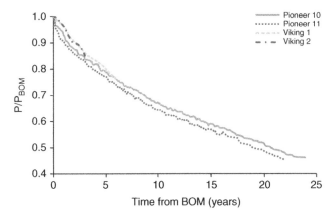

Figure 8.6 Lifetime performance of the SNAP-19 flight systems normalized to power at BOM. For the purposes of this graph, BOM for Viking is the point when the mission landed on Mars.

Figure 8.6 presents the available performance for the SNAP-19 missions. Performance data for the SNAP-19 RTGs on Pioneer and Viking were taken from Ref 23 and Ref 7, respectively. These are the first RTGs to demonstrate high reliability, stability, and extreme longevity. Power losses were slow, graceful, and allowed missions to operate for decades. It is important to mention that BOM is assessed differently for the Pioneer and Viking missions. The deep space Pioneer missions marked BOM at liftoff from Earth. The Viking missions marked BOM when the landers touched down on Mars. Thus, the Viking RTGs experienced some power losses during the cruise to Mars and the time spent in orbit looking for a landing site. Those performance losses are not shown in Figure 8.6. Ergo, one cannot draw too much meaning from the early life differences between the Viking and Pioneer RTGs.

Despite the strong similarities in RTG behavior for these four spacecraft, we can observe a few interesting things. First, the performance of the Pioneer 11 RTGs was slightly worse than Pioneer 10's. This is not surprising because Pioneer 11 was the backup probe, and all the best components (and RTGs) flew on Pioneer 10. [19] What is most interesting about this is that even with Pioneer 10 getting the four best performing RTGs, and 11 getting the four lowest performing generators, the difference in performance losses after 22 years is only 2.5%. This was the first time RTGs showed a high degree of reliability and stability, traits that define our expectation of modern RTG performance.

Another interesting observation is that the power losses in Viking appear to be more severe than Pioneer. Conditions on Mars are harder on an RTG than deep space. The very low-pressure atmosphere is a good insulator, and Mars's proximity to the sun means that solar heating can affect the generator. These effects cause temperatures on the hot side of the thermoelectric converter to be higher on Mars, and high temperatures can increase the rate of thermoelectric degradation. While the Martian environment appears to be a probable cause for increased power losses, it is also possible that one of the many design changes made to adapt RTGs to Mars could also contribute to these power losses.[23]

Finally, the performance of the RTGs on both Viking landers displays a long, wave-like pattern. The RTGs in both landers experienced this performance variance, which loosely tracks the seasonal changes in a Martian year. Temperature changes of the hot junction of thermoelectric couples, and hence an RTG's performance, are highly dependent on heat rejection to the ambient environment. These observations suggest that seasonal temperature changes on Mars were large enough to influence the temperatures and power output of the RTGs. This effect has also been observed on the MMRTG that is powering the Curiosity rover on Mars. [20,21]

Performance of the SNAP-19 RTGs proved so good that it led to another unexpected first in space exploration. Pioneer only required the SNAP-19 to have a design life of three years. Long enough to study Jupiter and download all the data. The complete success of Pioneer 10 "loosened the reins" a bit because Pioneer 11 no longer needed to fill in any mission gaps. Based on the Pioneer 10 mission's performance, confidence in the Pioneer 11 RTG performance was so high the mission team altered the trajectory of the spacecraft mid-flight so that besides Jupiter, the probe could also become the first spacecraft to encounter Saturn. When it reached Saturn, seven years after launch, SNAP-19 defied expectations again by providing 33% more power than was required for the mission.

Unfortunately, complete lifetime performance is not available for Pioneer 10 or Viking 2. Pioneer 10 is an interesting case. The long life of Pioneer 10 is a point of pride for NASA. Pioneer 10's mission length is cited frequently as over 30 years [19], but routine tracking and data processing ended after 25 years for budgetary reasons. [22] Officially, however, the probe was not shut down. Occasional contact with the spacecraft was made for training exercises, but the scientific mission, including monitoring RTG power, had ended. Teledyne Energy Systems, Inc. published the most complete set of Pioneer power data that has been found, and that contains all Pioneer data through 1995. [23] The missing Pioneer 10 data (1/1/1996 through 3/31/1997) is disappointing. However, given the stability and longevity of Pioneer 10, it is very unlikely this short 15-month period would show any changes in the power trends.

---

23 For example, changes to the gas management system as discussed with the SNAP-19B in Section 8.3.3.

The Viking 2 lander also lived longer than the power data shows. An inadvertent shut down of the lander occurred after three years on Mars. When the lander was revived, the RTG data was no longer being transmitted. The lander continued to operate for another six months. After that, the orbital relay for Viking 2 ran out of propellant and could not maintain contact with the lander. Similar to Pioneer 10, it is very unlikely that the performance over the missing six months would show any major changes in the power trends.

The stories behind the end of Pioneer 11 and Viking 1 are less involved. After 22 years and 7 months, Pioneer 11 no longer had enough power to operate any of the instruments on board, so the spacecraft was shut down. This is one of the few examples of a mission ending, albeit well beyond its required lifetime, because the RTG could no longer power the spacecraft. Meanwhile, contact with Viking 1 was lost after six years of operations on Mars when the mission control team sent faulty commands to activate the battery charging cycle. [7]

8.3.7 Multi-Hundred Watt RTG

Momentum for the Planetary Grand Tour was building, and the AEC knew that SNAP-19 would not deliver enough power to meet this challenge. In the words of Bob Carpenter, "These missions require on the order of ten times as much power for mission lifetimes up to twice as long as any other mission to date with very tight constraints on size and weight to explore a region of space where we have never been before." [9] In order to meet these challenges, work began on a silicon germanium (SiGe) "Multi-Hundred Watt" RTG (MHW-RTG). SiGe thermoelectric couples can operate at much higher temperatures in a vacuum. This allowed the design to use the insulating properties of the vacuum of space to shed a lot of insulation mass. While the MHW-RTG conversion efficiency of 6.6% was marginally better than the SNAP-19 (6.2%), the real advantage to NASA was the 30% decrease in mass per watt of power. Since these missions had much larger power needs, this improvement in specific power (W_e kg^{-1}) resulted in a mass savings of almost 100 kg. This mass could then be allocated to other critical spacecraft systems (thruster propellant, scientific instruments, etc.).

NASA wasn't the only customer interested in more power. The DOD also approached the AEC with interest in an RTG system that could produce over a hundred watts. The combined interest and investments by NASA and the DOD enabled the development of MHW-RTG. Eventually, MHW-RTG would be launched on a series of marquee missions. A pair of experimental DOD communication satellites known as the Lincoln Experimental Satellites (LES) 8 and 9, and the two spacecraft that would complete the Planetary Grand Tour, Voyager.

Figure 8.7 presents the available data for MHW-RTG missions through mid-2021. Performance data for the Voyager and LES missions were supplied by the JPL

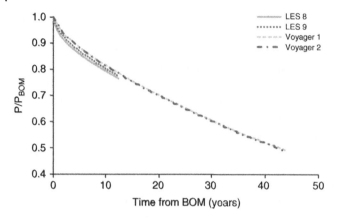

Figure 8.7 Lifetime performance of the MHW-RTG flight systems normalized to power at BOM.

mission control center and Ref 24, respectively. The most notable things about this data are the exceptional longevity and power losses that are much slower than SNAP-19. It is these combined achievements in longevity, stability (i.e., smaller power losses), and decreased mass that cause us to consider development of the MHW-RTG to be the most revolutionary advancement in RTG performance.

A lot of credit for longevity also needs to be given to the missions and the spacecraft. The history of RTGs is rife with examples of "losing communication with the spacecraft," or budget cuts that effectively, or literally, shut down the mission. It was the phenomenal feats of engineering in both spacecraft and RTG design that not only allowed Voyager to succeed at its very challenging mission, but for it to continue defying expectations as it exited the solar system and continues to explore interstellar space 44 years later.

Improved stability (or reduced power loss) in the MHW-RTG is because of the SiGe material. Degradation in a SiGe thermoelectric converter, compared to PbTe/TAGS, is both slower and less severe. In fact, after only 10 years, degradation in SiGe performance approaches the rate expected from thermal inventory losses (~1.2% per year). Total power losses in Pioneer's SNAP-19s after 25 years were still over two-times higher than thermal inventory losses. Since thermal inventory losses are tied to radioactive decay of the fuel, thermal inventory losses are a fundamental minimum amount of degradation that will occur in every RTG.

The fact that degradation in SiGe thermoelectric materials approaches this fundamental minimum degradation rate after such a short time is an extremely valuable trait for long-term missions. PbTe/TAGS start out with almost the same conversion efficiency, but their converter degradation accounts for almost half of the power losses. Copper selenide is a thermoelectric material that has a much higher conversion efficiency (8-9%), but attempts to mature the technology showed that degradation rates were so high that SiGe would provide more power after only a few years. Future RTG technologies will have to fill some big shoes. They will need to exceed the conversion efficiency of SiGe (6.6%) and have very low degradation in the thermoelectric material. This is a difficult combination to find and is the reason SiGe is still considered state-of-the-art for space RTGs over 44 years after they were sent on their first mission.

Unfortunately, complete performance data for the LES missions was not available. This is disappointing because the LES and Voyager spacecraft were operated in very different environments. As an Earth satellite, LES would experience average temperatures that were higher than Voyager. Estimates from Pioneer show that RTG temperatures at Earth are about 10 °C higher than in deep space. [12] LES also would have experienced a lot of temperature cycles as the satellite passed in and out of Earth's shadow. If there were differences in the performance of the LES and Voyager spacecraft, they may have told a more detailed story about how these stresses can affect a SiGe thermoelectric converter.

Despite the limited data from LES, a small difference in performance is observed between LES and Voyager power numbers. This is most likely because of the stresses caused by the higher temperatures and temperature cycles discussed previously. Higher temperatures will increase thermoelectric degradation. 10 °C should not cause a large increase in degradation rate, but after 12+ years, the power losses would likely be noticeable. In addition, Carnot efficiency will be lower in Earth's orbit because of the much higher heat sink temperature. Estimates from Pioneer suggest a 1% increase in power from Carnot efficiency improvements for the SNAP-19 when it reached Jupiter. [12] While this is not a large difference in power, losses because of operating at a lower Carnot efficiency could add up. Separately, the constant temperature swings experienced by an Earth satellite may have generated some mechanical stresses within the RTG. Those stresses may have led to lower power production. The effect of these thermal cycles on a SiGe converter has not been studied, so the potential impact on performance is unknown.

LES 8 was decommissioned after 28 years in orbit because the command system had become intermittent. LES 9 was decommissioned because of the loss of a primary telemetry system and a gradual reduction in RTG power. Thus, LES 9 is one of the few examples where RTG power losses contributed to termination of a mission. After 44 years of successful operation however, LES 9 is a paragon of RTG performance.

The MHW-RTGs on both Voyager missions continue to provide enough power to conduct meaningful science experiments 44 years after launch. Estimates from the mission suggest that the RTGs will continue to provide enough power until at least 2025, 48 years after launch. Don't hold us to that number, though. Voyager has spent its life defying expectations.

8.3.8   General Purpose Heat Source RTG

In the 1970s, a revolution was taking place in the materials science world. Three-dimensional carbon/carbon composites were being developed that could maintain their properties at extremely high temperatures, and could even endure these temperatures for a short time in air. For the first time, a low density, high thermal conductivity, robust material designed to endure re-entry was available. The specific carbon/carbon composite chosen for use in RTGs was Fine Weaved Pierced™ Fabric (FWPF). The improved re-entry and impact properties of FWPF provided an opportunity to enhance safety and simplify the scope of the systems used to contain plutonium during the worst spacecraft launch accident conditions. In addition, the low density and good thermal conductivity provided an opportunity to reduce mass and improve specific power. Thus, the AEC began "the most technically ambitious and challenging development program it had ever undertaken," [25] development of the General Purpose Heat Source (GPHS). The GPHS continues to be the only flight qualified design approved by the DOE for use in RTGs. It is easy to take the GPHS module for granted. It appears to be a simple little block of carbon.

That RTG historians and luminaries consider the GPHS module, the primary safety envelope for the fuel, to be the most technologically challenging development they had ever pursued, speaks volumes about the challenges of intact heat source recovery. And volumes about how important safety is to the DOE and NASA.

In terms of power demand, Voyager was only the beginning. RTG missions were getting bigger, more ambitious, and they needed a lot more power. Increased demand for power and development of the GPHS module led to the GPHS-RTG.[24] This new design would produce twice the power of MHW and incorporated both the successful SiGe converter and the improved GPHS safety system. This new design reduced the mass per watt of electricity by almost 20% over MHW.

From a performance perspective, specific power is the only area where the GPHS-RTG improved over the MHW- RTG. Since the SiGe converter is nearly identical in both designs, the conversion efficiency for GPHS-RTG remained at 6.6%. It also means that the reliability, stability, and longevity were the same. This modest improvement in specific power, coupled with the exceptional SiGe performance, and preservation of the robust design heritage of the MHW-RTG is why many consider the GPHS-RTG to be the best performing RTG ever produced.

Figure 8.8 presents the performance of GPHS-RTG missions through 2020. Performance data for Galileo and Cassini were supplied by the JPL mission

---

24 In order to reduce confusion, we will always refer to the RTG design as GPHS-RTG. When referring to just the module, we will always just use GPHS.

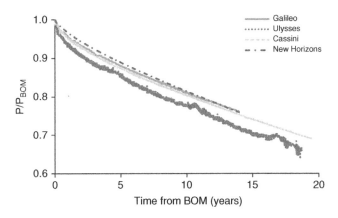

Figure 8.8 Lifetime performance of the GPHS-RTG flight systems normalized to power at BOM.

control center, New Horizons's data were supplied by the APL mission control center, and Ulysses data were supplied by the Mission Operations Manager, Nigel Angold. [31] Data in this figure is very similar to Figure 8.7, confirming the similarity to MHW in reliability, stability, and longevity. Recapturing the MHW-RTG state-of-the-art performance, despite major changes to the heat source and thermal design, was an impressive feat of engineering.

What is perhaps most interesting is that all the spacecraft that flew GPHS-RTG show similar performance, despite fairly different conditions surrounding their launch. Cassini was the only spacecraft to fly RTGs that were fueled and launched under nominal conditions. Galileo and Ulysses were both fueled in 1985, but the unfortunate Challenger disaster caused launch delays of about five years. [26] Fuel production issues caused New Horizons to be launched with some ^{238}Pu that was over 20 years old and resulted in a fuel loading (i.e., total heat) that was ~10% below normal. [27] Despite launching these spacecraft with RTGs at various stages of age and fuel loading, they all appear to show similar in-flight performance. RTG performance on the Ulysses spacecraft may be an exception, but the exception is suspect.

Looking at individual GPHS-RTG powered spacecraft shows a few interesting nuances. First, power losses on Ulysses appear to be noticeably higher than on the others. This nuance should be taken with a grain of salt. Ulysses is the only spacecraft presented in Figure 8.1 where power was not directly measured from the RTG's voltage and current. Power was calculated using an algorithm that included 1) main bus current, 2) internal power dump current, and 3) nominal power consumption from spacecraft components. [28] The exact nature of this algorithm is unknown. It is also likely that the algorithm was based on spacecraft properties before launch, and does not include the effects of age or burn-in[25] on the electronics. Considering the behavior of Ulysses is slightly out of family with every other GPHS-RTG and MHW-RTG powered mission, it seems more likely that the odd performance is due to flaws in the algorithm. Another interesting feature of the Ulysses data are the increases in power that occurred during the three solar encounters.[26] While the increased temperature caused by proximity to the sun created these "bumps," it is difficult to say if these represent real increases in power. It is also possible that the increased temperature affected some of the electronic components on the spacecraft (e.g., resistances), which then affected how the algorithm calculated power.

The New Horizons RTG also displayed some interesting nuances in behavior. During the first few years, the power losses were slightly smaller than any other MHW or GPHS-RTG. This makes sense because the thermal inventory on New Horizons was ~10% below nominal, which would produce lower temperatures, reduce the rate of thermoelectric degradation, and reduce power losses. Despite the reduced short-term degradation, long-term performance losses in New Horizons' RTG appear to be faster than Cassini and Galileo. A rate law analysis (see Chapter 7) shows that this is because the thermal inventory losses on New Horizons are higher. Thermal inventory losses on New Horizons are 1.47% per year, while a nominal mission like Cassini has thermal inventory losses of 1.24% per year. Interestingly, the proportional increase in thermal inventory losses for New Horizons is the same as the proportional decrease in thermal inventory in the RTG. In other words, the lower BOM power for New Horizons makes the percentage decrease from thermal inventory losses appear larger. When in fact, the decrease in watts of electricity caused by thermal inventory losses on New Horizons and Cassini are the same.

While the performance of Galileo's RTGs appear to be normal, there have been some errors in the public record regarding the performance of Galileo. Several publicly available sources cite the end-of-mission power for Galileo as 485 W_e, which would be a $P/P_{BOM} = 0.85$. It seems clear from the context of these references that "end-of-mission" is referring to the power produced by Galileo when it was decommissioned.[27] Figure 8.8 clearly contradicts this. The value of 485 W_e most likely comes from the power at the end of the primary mission, 8.2 years after launch. Galileo had two extended missions, however, and the spacecraft operated for nearly 14 years. Several sources have attempted to calculate power

---

25 A term used to describe the natural tendency for the electrical properties of components to change as they are used.

26 The three solar encounters occurred roughly at 5, 11, and 17 years.

27 Galileo was decommissioned by commanding it to enter Jupiter's atmosphere.

losses over time using 485 W_e, but applying a lifetime of 14 years. This error most likely originates from the *Galileo End-of-Mission Press Kit*, which contains the end-of-mission power typo. [29]

Thus, to set the help set the record straight, the power produced by Galileo's RTGs when it was decommissioned is reported here as 438 W_e. That value is taken from Galileo spacecraft data, which were supplied by JPL, the mission control center for Galileo. Other available documents cite an end-of-mission power of 432 W_e, but they also cite a BOM power of only 570 W_e. [30] Lockheed Martin,[28] however, cites the RTG power for Galileo as 576 W_e at BOM. [24] Given their pedigree, the authors are inclined to believe the spacecraft data (438 W_e at decommissioning) and Lockheed Martin report (576 W_e at BOM) provide a more accurate description of RTG power.

Galileo was decommissioned because of extensive damage caused by flights through Jupiter's trapped radiation fields, raising the possibility that loss of control of the spacecraft could allow the spacecraft to crash into, and potentially contaminate, one of Jupiter's moons. New Horizons continues to explore the Kuiper Belt in the outer reaches of the solar system, and is expected to have enough power to continue operating late into the 2030s. Cassini was decommissioned because it ran out of thruster propellant, which caused the spacecraft to pose a risk to planetary protection. Saturn's moons, Titan and Enceladus, have environments that could support life, or the molecular precursors necessary for life. Similar to Galileo, without adequate thruster propellant, Cassini would no longer be in control of its flight path, making it possible to contaminate one of Saturn's moons.

Communications with Ulysses were lost in 2009. We believe the end-of-mission was caused by a combination of a failing communication system and a reduction in heat and electrical power from the RTG. RTG thermal output had decreased enough that the hydrazine thruster propellant may have frozen, and thrusters were used to point the spacecraft towards Earth for communications. The durability of the spacecraft and RTG, however, allowed it to complete three full solar encounters over 19 years before it went dark. The original mission only called for a single solar encounter.

8.3.9 Multi-Mission RTG

NASA considered an early design concept for the Viking landers that used rovers. Static, RTG-powered landers were ultimately selected for the Viking missions. Decades later, in 2003, NASA was planning to send another rover to

---

28 Prime contractor for GPHS-RTG and system integrator for Galileo.

Mars. All rovers previously landed on Mars by NASA[29] received their electrical power from solar panels. The 2003 conceptual rover was larger and was to carry a much more ambitious set of science experiments than the solar powered rovers. In addition, the rover was to land, survive, operate, and last for two Martian years within a latitude band of +/− 30 degrees about the equator. The band imposed extreme weather on the rover that would be compounded by having to survive night times. Both would consume energy for heat, and only operating this flagship mission during the daytime on Mars was an unacceptable limitation. An RTG quickly became the obvious solution. It could provide continuous power day or night, and waste heat could be scavenged to warm much of the rover.

SiGe RTGs are open to their environment—they rely on the vacuum of space for insulation. That design feature is incompatible with Mars. Carbon dioxide in the low pressure Martian atmosphere would rapidly react with sensitive components inside a SiGe RTG, such as the GPHS modules and the molybdenum foil insulation. [32] In order to ensure long-term operation on Mars, the RTG had to be hermetically sealed. Thus, NASA commissioned a new RTG based on the old PbTe/TAGS technology that could operate under a cover gas; an RTG derived from the ones used on the Viking spacecraft.

While a SNAP-19 RTG had not been produced in almost 30 years, the PbTe/TAGS thermoelectric technology had been kept alive by other government programs. This allowed rapid development of the Multi-Mission RTG (MMRTG). An RTG based on SNAP-19 with several modest design improvements based on lessons learned over the past three decades, NASA's general desire for more power, and the need to operate on Mars. Conversion efficiency for the MMRTG is equal to SNAP-19 (6.2%), and specific power is equal to the Viking design (2.8 W_e kg^{-1}).

Figure 8.9 presents the performance of the MMRTG flight system on the Mars Curiosity rover. Performance data for Curiosity were supplied by Aerojet Rocketdyne, the prime contractor for the MMRTGs. As discussed previously, BOM for most spacecraft is defined as the time of launch from Earth. In order to remain consistent with other missions to Mars (e.g., Viking) and the literature on Curiosity, however, the BOM for Curiosity (Figure 8.9) will be referenced to landing on Mars.

The MMRTG power data are suggestive of a reliable and long-lived RTG similar to the Pioneer mission RTGs (Figure 8.2). Power degradation of the MMRTG on the Curiosity rover is less than what was observed for the Pioneer RTGs. After 7.5 years, the MMRTG is producing a small (3%), but noticeable, improvement in efficiency over Pioneer 10 at the same point in time. This is notable because

---

29 Sojourner, Spirit, and Opportunity.

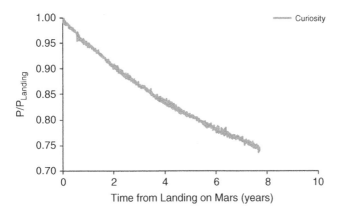

Figure 8.9 Lifetime performance of the MMRTG flight system powering Curiosity normalized to the power at the landing on Mars.

environmental conditions on Mars are harder on an RTG than a deep space cruise, like the two Pioneer flights experienced. Experiencing less degradation for an MMRTG on Mars suggests that the modest design improvements in the MMRTG were successful.

Overall, MMRTG performance appears to be in-family with the SNAP-19. As discussed earlier in this chapter, this is because lifetime performance is driven by the type of thermoelectric converter. RTGs that use PbTe/TAGS thermoelectric couples are expected to have similar performance because they degrade by the same mechanisms and operate at similar hot junction temperatures. Increased power in the Curiosity MMRTG is a modest, and welcome, improvement in performance. This improvement is due to optimizing the design, not major improvements in the degradation mechanisms.

Results from life testing the MMRTG Engineering Unit confirm these observations on reliability, stability, and longevity. [21,33,34] This improves confidence that the behavior of the MMRTG powering Curiosity will represent future MMRTG flight systems.

In terms of expected lifetime, the success of SNAP-19 suggests that the MMRTG could outlive its 17-year design life and operate into the 2030s.

The Perseverance rover was also launched with an MMRTG. Perseverance had only been on Mars for seven months at the time this chapter was written, and the MMRTG was performing as expected at that time. Seven months is not enough time to draw any meaningful conclusions about the RTG's performance, so Perseverance data is not in Figures 8.1 or 8.9.

The Dragonfly mission designers have chosen to fly an MMRTG and plan to land with it on the moon Titan and operate their rotorcraft in the thick atmosphere. A hermetically sealed MMRTG is the best choice for a mission to this moon of Saturn. As discussed previously, SiGe-based RTGs are open to their environment because they rely on the vacuum of space for insulation. The dense atmosphere of Titan would diffuse into a SiGe-based RTG and the gases would carry some of the RTG heat away from the thermoelectric converter. That would lower temperatures and significantly reduce power. Plus, the high temperatures of a SiGe-based RTG would cause parts of the Titan atmosphere to undergo chemical reactions and create undesirable products, such as ammonia. [32]

References

1 Hammel, T., Otting, B., Bennett, R., and Sievers, B. (2015). RTG degradation primer and application to the MMRTG. *Proceedings of the Nuclear and Emerging Technologies for Space (NETS 2015)*, Albuquerque, NM (February 2015), 5107. American Nuclear Society.

2 VanderVeer, J.R. (2017). Development of a high performance multi-physics finite difference model for use in a monte carlo simulation with real world distributions. *Proceedings of the Thermal and Fluids Engineering Conference (TFEC 2017)*, Las Vegas, NV (April 2017), TFEC-IWHT2017-17687. https://doi.org/10.1615/TFEC2017.cfd.017687

3 Wood, E.G., Herman, J.A., Hall, R.A. et al. (2016). Multi-mission radioisotope thermoelectric generator experience on Mars. *Proceedings of the Nuclear and Emerging Technologies for Space (NETS 2015)*, Huntsville, AL (February 2016), 6024. American Nuclear Society.

4 Stapfer, G. (1976). Degradation Model for an RTG with a Silicon-Germanium Thermopile. *DOE/ET/33003-T4*. Jet Propulsion Laboratory, California Institute of Technology, Pasadena, CA, https://doi.org/10.2172/6164169

5 Whiting, C.E. (2020). Proposed standard for extrapolating limited RTG performance data to establish behavior trends and a lifetime performance prediction. *Proceedings of the Nuclear and Emerging Technologies for Space (NETS 2020)*, Knoxville, TN (April 2020). Oak Ridge National Laboratory.

6 Anon (1969). TAGS-85/2N RTG Power for Viking Lander Capsule. *INSD-2650-29*. Teledyne Isotopes, Nuclear Systems Division, Baltimore, MD. https://doi.org/10.2172/5415537.

7 Bennett, G.L., Lombardo, J.L., and Rock, B.L. (1984). US radioisotope thermoelectric generators in space. *Nucl. Eng.* 25 (2): 49–58.

8 Anon (1978). Artificial Earth Satellites Designed and Fabricated by The Johns Hopkins University Applied Physics Laboratory. *SDO-1600 (Revised)*. The Johns Hopkins University Applied Physics Laboratory, Space Department, Laurel, MD.

9 Anon (1987). Atomic Power in Space. *DOE/NE/32117-H1*. US Department of Energy, Washington, DC. ISBN: B00161GGKA

10 Carpenter, R.T. (1971). Space nuclear power systems. *Proceedings of the National Symposium on Natural and Manmade Radiation*, Las Vegas, NV (March 1971). https://doi.org/10.2172/4624408

11 Fihelly, A.W. and Baxter, C.F. (1970). The SNAP-19 radioisotope thermoelectric generator experiment. *IEEE Trans. Geosci. Electron.* 8 (4): 255–264. https://doi.org/10.1109/TGE.1970.271419.

12 Anon (1973). SNAP 19 Pioneer F & G Final Report. *DOE/ET/13512-T1*. Teledyne Isotopes, Energy Systems Division, Baltimore, MD. https://doi.org/10.2172/5352675

13 Barklay, C., Whiting, C., Schmitz, P., and Sutliff, T. (2021). Can MMRTG operate on the Moon? Insights from SNAP-27 for Apollo Lunar surface experiments package. *Proceedings under IEEE Conference on Aerospace,* Big Sky, MT (March 2021). IEEE Publications. https://doi.org/10.1109/AERO50100.2021.9438297

14 Bennett, G.L., Hemler, R.J., and Schock, A. (1996). Status report on the US space nuclear program. *Acta Astronaut.* 38 (4-8): 551–560. https://doi.org/10.1016/0094-5765(96)00038-0.

15 Black, J. (1967). Thermoelectric Leg Product Specification (TELPS), Final Technical Report, Volume 1, Part 2. *6300-262*. General Electric Co., Philadelphia, PA. https://doi.org/10.2172/4509687.

16 Hammel, T. (2021). *Personal Communication*

17 Launius, R.D. (2004). *Frontiers of Space Exploration*, 2e. Westport, CT: Greenwood Press. ISBN: 978-0313325243.

18 Uri, J. (2019). 40 years ago: pioneer 11 first to explore saturn. In: *NASA History*. FL: NASA Johnson Space Center www.nasa.gov/feature/40-years-ago-pioneer-11-first-to-explore-saturn (accessed 30 November 2021).

19 Anon (2003). Last signal sent from RTG-powered spacecraft. *Nuclear News* (April), p. 65–67. http://www2.ans.org/pubs/magazines/nn/docs/2003-4-3.pdf

20 Whiting, C.E., Kramer, D.P., and Barklay, C.D. (2019). Empirical power prediction for MMRTG F1. *Proceedings of the Nuclear and Emerging Technologies for Space (NETS 2019)*, Richland, WA (February 2019). American Nuclear Society.

21 Whiting, C.E. (2020). Understanding the MMRTG Lifetime Performance Using Modeling and Analysis. *APS4DS 2020*. Pasadena, CA.

22 Furlong, R.R. and Wahlquist, E.J. (1999). US space missions using radioisotope power systems," *Nuclear News* (April), p. 26–34.

23 Hammel, T. (1996). Power history of pioneer 10 and 11 RTGs. In: *Personal Archives*. Graph on Teledyne Energy Systems Letterhead.

24 Anon (1998). GPHS-RTGs In Support of the Cassini RTG Program – Final Technical Report. *DOE/SF/18852-T97*. Lockheed Martin Astronautics, Philadelphia, PA. https://doi.org/10.2172/296824

25 Carpenter, R.T. (2012). *Personal Communication*

26 Bennett, G.L., Lombardo, J.J., Hemler, R.J., and Peterson, J.R. (1986) The general-purpose heat source radioisotope thermoelectric generator: power for the Galileo and Ulysses missions. *Proceedings of the 21st Intersociety Energy Conversion Engineering Conference*, San Diego, CA (August 1986). American Chemical Society.

27 Fountain, G.H., Kusnierkiewicz, D.Y., Hersman, C.B. et al. (2009). The new horizons spacecraft. In: *New Horizons* (ed. C.T. Russell), 23–47. New York, NY: Springer. ISBN: 978-0387895178.

28 Bennett, G.L., Hemler, R.J., and Shock, A. (1994). Development and use of the Galileo and Ulysses power sources. *45th Congress of the International Astronautical Federation*, Jerusalem, Israel, IAF-94-R.1.362. https://doi.org/10.2172/1033366

29 Anon (2003). *Galileo End of Mission Press Kit*. Pasadena, CA: National Aeronautics and Space Administration/Jet Propulsion Laboratory-Caltech.

30 Anon (2003). Galileo mission to Jupiter. In: *NASA Facts*. Pasadena, CA: National Aeronautics and Space Administration/Jet Propulsion Laboratory-Caltech.

31 Angold, N. (2014). *Personal Communication*

32 Whiting, C.E., Marth, J.L.D., and Barklay, C.D. (2019). Effect of Martian and Titan atmospheres on carbon components in the general purpose heat source. *Proceedings under IEEE Conference on Aerospace 2019*, Big Sky, MT (March 2019), 2069. IEEE Publications. https://doi.org/10.1109/AERO.2019.8741909

33 Whiting, C.E. (2020). Empirical performance analysis of MMRTG power production and decay. *Proceedings under IEEE Conference on Aerospace 2020*, Big Sky, MT (March 2020), 2060. IEEE Publications. https://doi.org/10.1109/AERO47225.2020.9172270

34 Whiting, C.E., Barklay, C.D., and Tolson, B.A. (2022). Understanding Degradation and Heat Losses in the MMRTG and an Improved Analysis of F1 on Curiosity. *Proceedings under IEEE Conference on Aerospace 2022*, Big Sky, MT (March 2022), 2055. IEEE Publications (2022).

9

Modern Analysis Tools and Techniques for RTGs

Christofer E. Whiting^a, Michael B.R. Smith^b, and Thierry Caillat^c

^{a} *University of Dayton Research Institute, Dayton, Ohio*
^{b} *Oak Ridge National Laboratory, Oak Ridge, Tennessee*
^{c} *Jet Propulsion Laboratory/California Institute of Technology, Pasadena, California*

RTGs are sophisticated and unique power sources for deep space science missions. These systems are subject to launch vibrations, extreme thermal cycling, and exposure to radiation environments. Furthermore, each RTG must operate during its maintenance-free lifetime of decades in space. These demanding performance requirements and resultant engineering complexities demand the utmost confidence in understanding an RTG's lifetime behavior, system performance, reliability, and associated operational phenomena.

To meet this need, computational modeling tools and analytical methods to evaluate and predict RTG performance have been developed over several decades. This chapter explores some of the modern modeling tools and analytical methods used to understand various phenomena associated with RTGs. These tools and methods predict thermoelectric couple physics, generator performance, thermal management, ionizing radiation effects, and more. While the topics presented here represent important capabilities of modern RTG modeling techniques, they represent only a subset of the vast array of tools and methods available for modeling these complex systems.

9.1 Analytical Tools for Evaluating Performance Degradation and Extrapolating Future Power

Analysis of *in situ* telemetry is important for NASA mission planning as well as evaluating systems under development, such as the US Next Gen RTG [1] and the European ^{241}Am based RTG. [2] This information will help NASA understand

The Technology of Discovery: Radioisotope Thermoelectric Generators and Thermoelectric Technologies for Space Exploration, First Edition. Edited by David Friedrich Woerner.

RTGs as an energy conversion technology, characterize the behavior of specific designs, and assess performance of individual units. Detailed analyses of flight telemetry from RTGs have not been performed because power degradation occurs through several complex mechanisms, some of which mask other mechanisms. These mechanisms are categorized under two major phenomena: 1) decreasing radiogenic heat, and 2) degradation of the thermoelectric converter efficiency.[1]

Heat loss from the natural decay of radioisotopic fuel affects several key operating parameters by changing the thermal power available for conversion, system temperatures, Carnot efficiency, ΔT across the thermoelectric materials, and integrated ZT of thermoelectric couples. Power degradation associated with these effects is additive, follows the decline in heat from decay of the radioisotope heat source, and is referred to as thermal inventory losses.[2]

Thermoelectric converter degradation is a more complex phenomenon that results from physical, chemical, and/or material changes in the converter. To complicate matters, some designs (e.g., SNAP-19) can experience multiple different thermoelectric degradation mechanisms if the RTG has a long operating life. Some known thermoelectric degradation mechanisms experienced by PbTe based RTGs include sublimation of PbTe, Fe poisoning from the iron hot-shoe, dopant diffusion, and degradation of the metallurgical bonds. [3–5] Known degradation mechanisms for SiGe converters include dopant precipitation, which generates changes in thermoelectric properties over time, sublimation, and changes in the thermal/electrical properties of the multi-foil insulation. [6–10]

A process based on chemical rate law equations can mathematically decompose and analyze *in situ* performance of an RTG. [11] Results show how this method can separate degradation mechanisms and facilitate mathematical analyses of degradation rates in individual RTGs. A derivation of the rate law equations and their use is provided here. This section provides context useful in understanding the performance of flown RTGs in Chapter 8, and helps users apply this method to analyze RTG telemetry.

9.1.1 Integrated Rate Law Equation

The key to the rate law method is to separate power degradation into its individual components and determine the integrated rate equation for each component. Integrated rate equations are obtained by applying the classical rate law equation to RTG performance:

$$Rate = \frac{-dP}{dt} = kP^x$$

---

1 As a reminder, the thermoelectric converter includes the thermoelectric couples and the insulation that helps direct heat flow through the couples.

2 Thermal inventory is the term used to describe the amount of thermal energy, or heat, produced by the fuel.

Where P is power, k is the rate constant, x is the order of reaction, and *Rate* is the instantaneous slope in the performance data. A common approximation for the instantaneous slope in real data is:

$$\frac{-dP}{dt} \approx \frac{-\Delta P}{\Delta t}$$

Where ΔP is the change in power over the time interval, Δt. Accuracy of this approximation increases as Δt gets smaller.

Rearranging the rate law:

$$-P^{-x}dP = k\,dt$$

And integrating from time t_1 to t_2:

$$\frac{-1}{(-x+1)}P_{t_2}^{(-x+1)} - \frac{-1}{(-x+1)}P_{t_1}^{(-x+1)} = kt_2 - kt_1$$

Rearranging and simplifying this result produces the integrated rate equation:

$$P_{t_2}^{(-x+1)} = k'(t_2 - t_1) + P_{t_1}^{(-x+1)}$$

Where k' includes all constants from the previous equation. Note there is a special case when $x = 1$, which causes the integrated rate equation to become:

$$\ln P_{t_2} = -k'(t_2 - t_1) + \ln P_{t_1}$$

A small error in x can have a disproportionate influence on the results from the integrated rate equation as x approaches 1. Better accuracy and precision are found by setting $x = 1$ in these cases. As a general rule, use this approximation when x is between 0.7 to 1.3. Throughout the rest of this text, the $x = 1$ case is assumed when appropriate.

9.1.2 Multiple Degradation Mechanisms

A key assumption required for the use of integrated rate equations is that the rate must be dominated by a single degradation mechanism.[3] Multiple mechanisms can be dominant for a period and then wane to be replaced by other dominant mechanisms. This can happen throughout the operating life of an RTG. One mechanism that influences degradation in all RTGs is thermal inventory losses. Other mechanisms are tied to the thermoelectric converter design. Figure 9.1

---

3 A single "mechanism" can be composed of multiple individual chemical, physical, or material changes. As long as each change has the same order of reaction (i.e., x), they are analyzed as if they are part of a single, multi-component mechanism. For example, $k_{TE_1}P^{2_{TE_1}} + k_{TE_2}P^{2_{TE_2}}$ simplifies down to $k_{TE_{1+2}}P^2$.

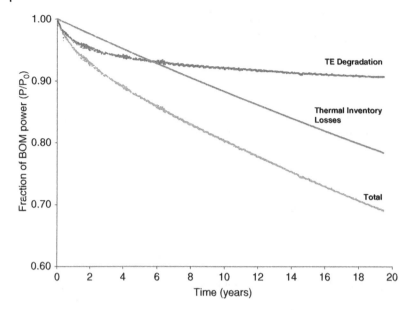

Figure 9.1 Power generation for Cassini RTGs as a fraction of the beginning-of-mission power. Contributions from thermal inventory losses and thermoelectric (TE) degradation are plotted for comparison.

presents an example from Cassini RTG telemetry and shows how changes in thermal inventory losses and thermoelectric converter degradation contributed to the overall performance.

When multiple mechanisms influence the rate in an RTG, the rate law equation becomes:

$$Rate = k_{Thermal}P^{x_{Thermal}} + k_{TE_1}P^{x_{TE_1}} + \ldots k_{TE_n}P^{x_{TE_n}}$$

Where $k_{Thermal}$ and $x_{Thermal}$ correlate to thermal inventory losses, and k_{TE_n} and x_{TE_n} correlate to each thermoelectric converter degradation mechanism.

Three methods can estimate or measure the effect of thermal inventory losses on RTG performance. First, changing the thermal input to an electrically heated RTG allows an empirical measurement of thermal inventory losses. Second, *a priori* computational techniques can estimate thermal inventory losses using models of RTG operation with different thermal inventories. Finally, an iterative numerical solution can be obtained by setting $x_{Thermal} = 1$,[4] using a reasonable estimate for

---

4 $x = 1$ is used because radioactive decay obeys an $x = 1$ mechanism, and radioactive decay is the fundamental process that drives thermal inventory losses. While the other components of the thermal inventory losses mechanism may not strictly follow $x = 1$, the fact that radioactive decay is the driving force makes the $x = 1$ approximation reasonable.

$k_{Thermal}$, subtracting the estimated thermal inventory losses from the rate, analyzing the remaining power losses, and then varying $k_{Thermal}$ until the best results are found. [11] The contribution of thermal inventory losses can be subtracted from the overall rate once they are known. The remaining performance profile will be from thermoelectric converter degradation.

When multiple thermoelectric degradation mechanisms are present, it is likely the rate will initially be controlled by the fastest mechanism (e.g., $k_{TE_1} P^{x_{TE_1}} \gg k_{TE_2} P^{x_{TE_2}}$). As the system ages, this fast mechanism will approach zero, and a different mechanism will dominate (e.g., $k_{TE_1} P^{x_{TE_1}} \ll k_{TE_2} P^{x_{TE_2}}$). These approximations show that when a single mechanism dominates the thermoelectric degradation rate, it is possible to describe the rate with a single $k_{TE} P^{x_{TE}}$ term, and that term can be converted into an integrated rate equation. Thus, in RTG systems with multiple thermoelectric degradation mechanisms, it is necessary to separate the data into components controlled by each mechanism and analyze them individually.

9.1.3 Solving for k' and x

Before the order of reaction (x) and adjusted rate constant (k') can be determined, it is necessary to reduce noise by smoothing the data. RTG power output is noisy because changes in the surrounding environment can influence power output. For example, power produced by the MMRTG on the Curiosity rover fluctuates due to Martian weather, seasons, and diurnal cycles. These weather phenomena can produce daily power fluctuations larger than 10 W_e or approximately 10% of average power. [12] Averaging data over large periods of time is one data smoothing technique that can reduce or eliminate noise in the data. Data averaging is the smoothing technique that will be used throughout this discussion.

Using averaged data, x is found by plotting log of the *Rate* vs. log of power. Sections of this log-log plot that produce a straight line are dominated by a single degradation mechanism, and a linear regression of that segment will have a slope equal to x. Figure 9.2 presents a log-log plot of the first five years of thermoelectric degradation on the Cassini spacecraft. [11] While Figure 9.2 still contains some noise, the trend observed in the log-log plot appears linear. This further suggests that the thermoelectric degradation observed on Cassini over the first five years was controlled by a single mechanism. A linear regression of this data shows that $x = 45.666$.

The log-log plot is also the reason why data smoothing is necessary. When using this rate law analysis, the smoothed power data must be constantly decreasing. If an increase occurs, a negative *Rate* will be produced, and the log of a negative number is undefined. Including undefined values in the analysis creates flawed results when linear regressions are applied to the data.

Curvature, or other non-linearity, in the log-log plot is a significant observation. One of two things may be happening if the data in a log-log plot is non-linear: 1)

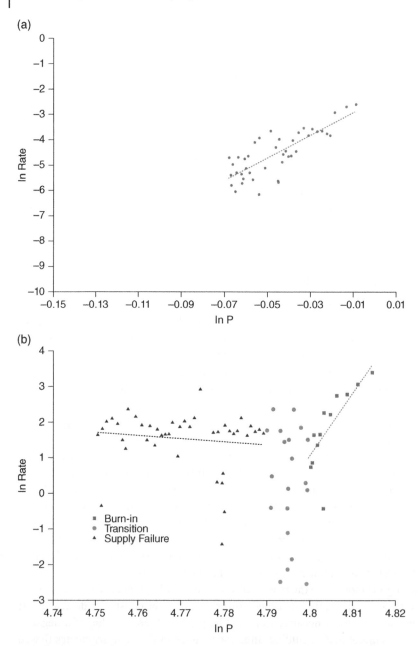

Figure 9.2 Example log-log plots of RTG data. (a) The first five years of Cassini data showing a linear trend over the entire data set. (b) The first 554 days of MMRTG Engineering Unit performance showing linear behavior when one of the two apparent mechanisms dominates, and a lack of linearity (i.e., curvature) during the transition between mechanisms.

$x_{Thermal}$ and/or $k_{Thermal}$ may not be correct, or 2) the smoothing process may not have produced a good representation of the data. When the log-log plot is linear in some areas, but non-linear in others, it shows that the dominant degradation mechanism is changing. Linear sections are dominated by one mechanism, while non-linear sections are where multiple mechanisms influence the rate. Figure 9.2 presents MMRTG Engineering Unit performance as an example of non-linear behavior. This plot shows that degradation was first dominated by a thermoelectric burn-in mechanism, followed by a transition period, followed by the slow failure of the heater power supply. [13]

A "rate constant plot" of $P^{(-x+1)}$ as a function of time can be produced after x has been found. A properly constructed rate constant plot should produce a straight line with a slope equal to k'. If x is wrong, the rate constant plot will be curved. If the degradation mechanism changes, the plot will curve where the mechanism is changing, and remain linear where a single mechanism is dominant. Thus, the curvature/linearity of the rate constant plot provides an additional self-check of analysis quality. In addition, observing slight curvature at the end of the rate constant plot could be an early indicator that the RTG behavior is changing.

Figure 9.3 presents an example rate constant plot produced using Cassini data with $x = 45.666$. A linear regression of this plot produces $k' = 3.77$. The linearity seen in Figure 9.3 suggests a single mechanism dominated RTG degradation over the 19.5-year Cassini mission.

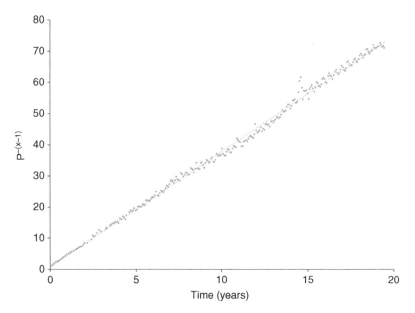

Figure 9.3 Rate constant plot for Cassini RTG performance after thermal inventory losses have been removed and $x = 45.666$.

9.1.4 Integrated Rate Equation

Once x and k' have been found, producing the integrated rate equation is a simple matter of inserting these variables into the equations from Section 9.1.1. For Cassini, the integrated rate equation for the thermoelectric degradation mechanism is:

$$P_{t_2}^{-44.666} = 3.77\left(t_2 - t_1\right) + P_{t_1}^{-44.666}$$

The linear nature of the log-log and rate constant plots shows that this integrated rate equation is valid for the entire 19.5-year Cassini mission. The contribution to RTG performance from thermoelectric converter degradation can be calculated using this equation for any point in time.

The integrated rate equation can also extrapolate the expected performance for a future date. It is very important to remember, however, that extrapolation assumes that the dominant thermoelectric degradation mechanism does not change. For some thermoelectric materials, like SiGe, this appears to be a good assumption. Other thermoelectric materials, like PbTe based converters, exhibit changes in degradation later in life. In these cases, extrapolating future performance should be performed with appropriate caution and caveats.

9.1.5 Analysis of Residuals

The result of a rate law analysis is an equation (i.e., the integrated rate equation) that describes the behavior of the data. An analysis of residuals, therefore, becomes one of the primary methods for assessing the quality of an integrated rate equation. Residuals are the difference between the actual data and the integrated rate equation results, with relative residuals being the residual value divided by the actual data.

Another self-check of analysis quality is the shape of a plot of residuals. Relative residual values that cluster around 0% suggest that $x_{Thermal}$, $k_{Thermal}$, and the integrated rate equation for thermoelectric degradation are accurate. Significant curvature in the relative residuals suggests that $x_{Thermal}$ and/or $k_{Thermal}$ are not accurate. Residual values that deviate from 0% could be a sign of inaccuracy in the integrated rate equation. Residual values near 0% for a period followed by ones that deviate from 0%, could indicate a change in the dominant degradation mechanism. A different integrated rate equation may be needed to describe this new behavior.

Figure 9.4 presents an example of residuals plotted for the Cassini mission, and it shows a tight distribution around 0% over the entire 19.5-year mission. This suggests that performance degradation of the thermoelectric converter during the Cassini mission was dominated by the same mechanism. Statistical analysis of the relative residuals shows the mean is −0.034% with a standard deviation of 0.082%.

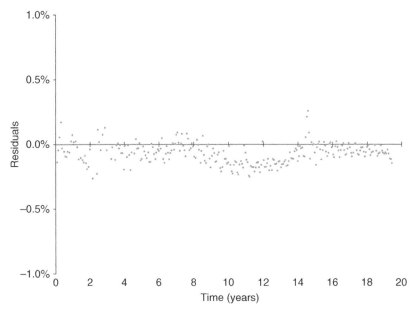

Figure 9.4 Plot of residuals for the thermoelectric converter degradation.

This suggests that the error in the integrated rate equation is less than 0.3% at any point during the Cassini mission ($3\sigma = 0.246\%$).

9.1.6 Rate Law Equations: RTGs versus Chemistry versus Math

In the field of Chemistry, rate law equations were derived with an emphasis on correlating the equations to collision theory, a theory that defines chemical reactions by their ability to interact, or "collide." According to collision theory, the order of reaction (x) is the number of elementary particles that need to interact simultaneously in order for the reaction to proceed, and the rate constant (k) is related to the probability that the molecules will interact.

In this work, we do not attempt to assign any special meaning to the values of x or k. The rate law equations are simply tools used to analyze RTG performance because of the long-standing success these equations have in assessing the mathematics behind systems that change. Currently, there is no known, or intended, correlation between the rate law constants and any chemical, physical, or materials phenomena.

Mathematically, the values of x and k define the type of curve that is being analyzed. x describes how much "curvature" exists in the curve. When $x = 0$, the result is a straight line, and when x gets very, very large, the curve starts to resemble a $90°$ angle. k, on the other hand, describes the slope of the curve. A larger k value will produce a larger drop in power (y-axis) over a specific time period (x-axis). Curves that have similar values of x will have similar degrees of curvature. Unfortunately, the corollary is not true for k because the value for k is dependent on x. In other words, two curves with different x values and the same drop in power over time will have different k values.

9.1.6.1 Application to RTG Performance

The method presented here was used to analyze performance of RTGs flown by the US. [14] Some of these missions did not live long enough to produce enough data to create a high confidence integrated rate equation (e.g., Viking), while others stopped transmitting data during the very early stages of the mission (e.g., Transit 5BN-1). Analyses of the longer lived missions have provided significant insight regarding the performance and behavior of RTGs.

9.2 Effects of Thermal Inventory on Lifetime Performance

One primary cause of decreasing RTG power is degradation in the thermoelectric converter. Thermoelectric degradation can be produced by a variety of different mechanisms, and sometimes the mechanism controlling the overall rate can change. [15] Most thermoelectric degradation mechanisms are thermally activated, meaning degradation will increase/decrease with temperature. This suggests that thermoelectric converter degradation would drop by lowering the temperatures within an RTG.

Fueling a generator with a lower thermal inventory is one way to lower temperatures and reduce thermoelectric degradation. Thermal inventory is the term used to describe the amount of thermal energy, or heat, produced by the fuel. Thermal inventory can be lowered by blending the ^{238}Pu fuel with some ^{239}Pu, a long half-life isotope that does not contribute to thermal power. Leaving the fuel in storage also lowers ^{238}Pu content as it decays with a half-life of 87.7 years. This results in 0.787% reduction in thermal inventory each year from radioactive decay.

While lowering thermal inventory may reduce degradation rates, it will also reduce the RTG's beginning-of-life power. Relationships between thermal inventory, beginning-of-life power, and thermoelectric degradation are often complex. Sometimes, a reduction in thermal inventory may decrease the degradation rate, but the overall benefit to performance is offset by a reduction in beginning-of-life power. In other systems, a reduction in thermal inventory would reduce beginning-of-life power, but the reduced degradation would produce more power later in life. This latter case suggests there is an optimal thermal inventory for an established RTG design that would maximize power for a mission.

Unfortunately, testing RTGs to determine an optimal thermal inventory for all missions is not practical. A design of experiments would require $10s of millions of dollars' worth of RTGs to be put on test for about two decades. Besides the financial burden, it is likely the results would not be ready in time to support a mission. Thus, the most realistic means of determining the optimal thermal inventory is through analysis and modeling. This requires knowledge of the beginning-of-life power and the physics of thermoelectric converter degradation.

An analytical study of thermal inventory versus RTG performance is performed here. The analyses draw on data from SiGe based GPHS-RTGs and PbTe/TAGS based MMRTGs. These are emphasized for two reasons. First, data required for the analyses are available. Second, these two RTGs are ready for flight today, or similar to designs that are likely to be available in the future. This analysis shows that an intentionally reduced thermal inventory at the time of fueling may benefit MMRTG users, but will not benefit GPHS-RTG users.

9.2.1 Analysis of GPHS-RTG

Long-term performance data for GPHS-RTGs is available from 4 missions: Galileo, Ulysses, Cassini, and New Horizons. Delays and fuel production issues caused three of these missions to be launched with off-nominal thermal inventories. These differences make it possible to analyze the effect of thermal inventory on performance.

Cassini is often considered the marquee GPHS-RTG mission. In terms of power, it was the highest powered RTG mission launched to date with 3 GPHS-RTGs. It is considered the best representation of average GPHS-RTG performance because it flew the most units on a single mission. Finally, Cassini RTGs were assembled under nominal conditions and launched on schedule. This means that Cassini is the only GPHS-RTG mission expected to show completely nominal RTG performance. Thus, Cassini will be considered the baseline for GPHS-RTG performance.

Three GPHS-RTGs were assembled between September 1984 and June 1985 to support Galileo and Ulysses. [16] Both missions were scheduled for launch on the US space shuttle, but the unfortunate 1986 Challenger disaster caused both missions to be delayed. As a result, the 2 Galileo RTGs were held in storage for ~4.5 years, and the Ulysses RTG was held in storage for ~5.5 years.

GPHS-RTG storage conditions are considered benign and beneficial. They can reduce the unicouple hot junction temperature by over 275 °C (i.e., 725 °C or 1,337 °F). [17] This reduction in temperature should stop thermoelectric degradation in any RTG, and the safe hot junction temperature for GPHS-RTG is ≤ 750 °C (1,382 °F). This analysis, therefore, assumes that all reductions in BOM power for Galileo and Ulysses were from radioactive decay of the thermal inventory.

Table 9.1 Changes in Mission Power and Thermoelectric Degradation as a Function of BOM Thermal Inventory.

Mission	BOM Averages[1]			Averages After 14 Years[1]	
	Inventory (W$_{th}$)	Power (W$_e$)	ΔP from Cassini (W$_e$)	TE Power Losses (W$_e$)	TE and Inventory Losses (W$_e$)[2]
Cassini [19]	4396	296	0	26	26
Galileo [16, 19]	4256	288	8	22	30
Ulysses [16, 19]	4264	289	7	30	37
New Horizons [18]	3948	244	52	13	65

[1] Average of RTG performance from each mission, non-averages inferred for missions with 1 RTG.
[2] Includes TE losses and reduction at BOM compared to Cassini, does not include thermal inventory losses.

The New Horizons spacecraft was powered by 1 GPHS-RTG. Issues linked to ^{238}PuO$_2$ fuel processing prevented the mission from receiving a full allotment of fresh fuel. Several fuel pellets produced as back-up for the Galileo and Ulysses missions were required to fuel New Horizons on time for the scheduled launch. This 20-year-old fuel had undergone a significant amount of radioactive decay, which decreased the combined thermal inventory by ~10%. [18]

Table 9.1 presents the BOM power and thermal inventory for all 4 GPHS-RTG missions. Telemetry was analyzed using the rate law method described previously in Section 9.1. Power losses from thermoelectric degradation after 14 years of operation were obtained. Results from this analysis are also presented in Figure 9.5. [14] Fourteen years was the benchmark for comparison because the Galileo spacecraft was decommissioned after 14 years, so this limits the extent of comparable data.

The reduction in thermal inventory used in the Galileo and New Horizons RTGs was significant enough to reduce the thermoelectric couple degradation rate, with New Horizons showing a 50% decrease in thermoelectric converter losses. While the decrease in thermoelectric degradation is notable, the power gains from decreased degradation are offset by even larger losses in beginning-of-life power.

One hypothesis that could arise from this is that operating a GHPS-RTG at higher temperatures could produce more power. This is not a practical option because higher temperatures could cause some materials in the GPHS module to exceed their safety margins. These materials form the safety envelope that protects against

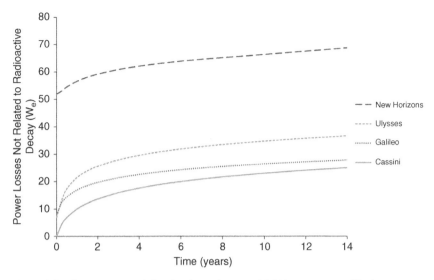

Figure 9.5 Power losses originating from decreased BOM power caused by lower thermal inventory, plus thermoelectric degradation losses for GPHS-RTG missions. Thermal inventory losses (i.e., power losses related to radioactive decay and heat loss) are not considered. Power at time = 0 uses the Cassini BOM power as a baseline (i.e., Power Losses = 0).

environmental dispersion of ^{238}Pu during a launch or re-entry incident. Increasing these temperatures is an unacceptable risk to environmental and human safety.

It is interesting that Ulysses launched with a thermal inventory similar to Galileo, but experienced a much higher rate of degradation. When Galileo's RTGs are compared to Cassini's, the five RTGs on those two missions show similar degradation rates. [19] The inconsistent degradation observed on the Ulysses mission is worth scrutiny. The mission flight trajectory does not explain the differences in degradation. Ulysses didn't experience any unusual thermal conditions until its first solar encounter almost four years after launch. A manufacturing flaw in the Ulysses RTG could cause unusual behavior. This is considered to be unlikely because of the reliable and robust performance of the other 6 GPHS-RTG flight generators on other spacecraft.

One probable explanation for the unusual telemetry is that Ulysses did not directly monitor RTG performance. Performance was estimated using an algorithm that included: 1) main bus current, 2) internal power dump current, and 3) nominal power consumption from spacecraft components. [20] None of the literature discusses the long-term accuracy of, or the potential effect of burn-in, on the algorithm. Burn-in is common for most electrical systems, and it refers to changes in the electrical properties of the system as it is used. Therefore, it seems likely that the unusual telemetry collected from Ulysses is not representative of the actual RTG performance. It is more likely that the unusual telemetry was caused by a minor flaw in or burn-in of the electrical components associated with, the algorithm.

These results suggest that while thermoelectric degradation in SiGe couples can be reduced, the losses from lower thermal inventory will always be larger. It also suggests that operation at the nominal 4410 W_{th} is the optimal GPHS-RTG configuration for any mission profile.

9.2.2 Analysis of MMRTG

A preliminary evaluation of the effect of thermal inventory on MMRTG performance can be estimated by analyzing life test data from the Engineering Unit (EU) and Qualification Unit (QU). The EU was placed on life-test with a nominal thermal inventory of 2000 W_{th}, and a fin root temperature (T_{fr}) of 182 °C (359 °F). [13] The QU was placed on life-test with a nominal thermal inventory of 1904 W_{th} and a T_{fr} of 154 °C (309 °F). [21]

Both units were electrically heated with a constant thermal input. A flight unit, on the other hand, experiences declining heat because of the radioactive decay of the ^{238}Pu fuel. Constant thermal input is an advantage for this analysis because there will be no thermal inventory losses. This makes the analysis simpler because all power losses are from thermoelectric converter degradation.

Unfortunately, data from the launched MMRTG flight units are not useful for this analysis. F1, powering the Curiosity mission, experienced a launch delay that caused the RTG to be kept in storage for three years. Unlike GPHS-RTG storage conditions, storage conditions for an MMRTG are not considered benign, resulting in some thermoelectric degradation during storage. While F2, powering the Perseverance mission, was assembled and launched under nominal conditions, the mission has been operating on Mars for less than a year at the time of writing. This is not enough data to draw any meaningful conclusions.

Before EU and QU results can be compared, it is necessary to make some minor adjustments. First, the QU was operated at a lower T_{fr}. This causes an increase in power that can be calculated using reference values found in the MMRTG User's Guide. [22] Second, when tested under similar conditions, the QU power output was 3.4 W_e lower than the EU. This difference is expected to be caused by normal unit-to-unit variations. QU data was adjusted to account for this lower beginning-of-life power, according to processes described in previous reports. [13, 21]

Data collected during life testing was analyzed using the rate law method described in Section 9.1, and the resulting rate law equations were used to extrapolate future performance. Total power losses were calculated by comparing these results against a beginning-of-life baseline value of 123.7 W_e. This was the average power produced by the EU during its first week of operations.

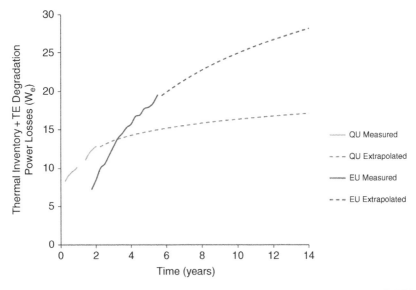

Figure 9.6   Power losses originating from thermoelectric degradation during MMRTG life testing plus decreased power from the lower QU thermal inventory. QU values account for expected differences in power output created by operating at a lower T_{fr}, and the measured 3.4 W_e reduction in power caused by unit-to-unit variations. Baseline power (i.e., Power Losses = 0) is set as the beginning-of-life power for the EU (i.e., 123.7 W_e).[5]

Figure 9.6 presents the total power losses produced by this analysis. These results show that the 4.8% decrease in thermal inventory for the QU reduced the beginning-of-life power by 7 W_e, but the rate of thermoelectric degradation was also reduced. The reduction in degradation is large enough that extrapolations predict the QU will outperform the EU after 3.5 years. It also suggests that operating at 1904 W_{th} would increase the end-of-design-life (14 years) power by 12.4 W_e, which is a 13% increase in power.

These preliminary results suggest missions whose highest power demand is at least three years after launch may benefit from using a thermal inventory of ~1900 W_{th}. Missions that need power less than three years after launch might benefit from the nominal 2000 W_{th} thermal inventory.

While these results seem promising, it should be reiterated that these results are preliminary. This is because the QU testing was halted after two years, so it could

---

5 NOTE REGARDING FIGURE 9.6 DATA: QU data from 0 to 90 days is not included because of acceptance testing and equilibrating with the test environment. QU data from 342 to 483 days is not included because the unit was being operated in off-nominal conditions to support the Perseverance mission. EU data from 0 to 554 days is not included because a slow heater power supply failure caused significant drift.

support the integration and launch of F2 on Perseverance. Additionally, rate law equations do not always fit early life data very well. [21] Visual inspection of the QU results also suggests some uncertainty in the extrapolation. Despite this uncertainty, the probability that the QU would outperform the EU seems high. It should also be noted that these extrapolations do not account for the late life change in mechanism that has been observed in the MMRTG. [15] While this late life mechanism change appears small, the effect is not understood empirically, so it cannot be accounted for in this analysis. These concerns can be evaluated in much greater detail if QU testing at 1904 W_{th} can be continued.

9.3 (Design) Life Performance Prediction

RTGs were developed in the 1950s and have powered several NASA spacecraft and have been key to numerous NASA deep space missions and scientific discoveries as summarized in Chapter 3. All the RTGs have used thermoelectric couples made from only two types of thermoelectric (TE) materials, silicon-germanium (SiGe) and lead-telluride (PbTe.) SiGe couples are used only in the generators for deep space missions. They are not viable on Mars or Titan or any other body with an atmosphere. Gasses within the atmospheres would contaminate the thermoelectric couples. Conversely, the MMRTGs include features that allow them to be used in the atmosphere of Mars or Titan. PbTe-based and TAGS TE materials have been used in the SNAP-19 generators and, more recently, in the MMRTG generators. SiGe-based TE materials have been used in the MHW and the GPHS-RTG generators. As these technologies and generators were developed, it quickly became critical to develop approaches and tools to predict the power for these generators as a function of time and operating conditions as long missions (possibly multiple decades) were contemplated. Two main approaches have been used for generating a time-dependent power prediction.

The first approach relies on previously flown (heritage) couples and generators and was, together with the Engineering Unit data as it became available, the main basis for the MSL F1 Multi-Mission Radioisotope Thermoelectric Generator (MMRTG) original time-dependent power prediction. [23, 24] While this approach has clearly the advantage of relying mostly on a "true" entire RTG degradation at the generator level for an environment, it can carry several limitations. One of which is the potentially limited availability of generator-level data across a wide range of environments, which can limit the ability of extrapolating/interpolating degradation across a broad range of mission environments. It can also be cost prohibitive to build and test enough generators to gain the data needed across a comprehensive range of conditions. This approach was used for the original MSL F1 prediction, which turned out to be overly optimistic. [24] Some changes in the couple design were implemented in the MMRTG couples compared to couples in the SNAP-19 and the degradation physics of the MMRTG couples was originally not sufficiently understood and therefore the SNAP-19 heritage data were not fully applicable. Since then, a significant amount of flight data has been generated from the F1 unit on Mars and data from the EU and the degradation model was updated and various approaches have been used to provide a longer-term power prediction although there remains some uncertainty for $\geq$ 17-year predictions. This first type of RTG life performance prediction model is often referred to as the heritage model.

The development of SiGe-based RTGs started in the mid-1970s and focused initially on the MHW-RTG. A new approach for predicting power in different environments was developed. [25] It relied on characterizing the degradation mechanisms for the SiGe couples and insulation around the couples as a function of time and temperature and developing a physics-based model for long-term power prediction. The original computer code used to model the generator performance was called DEGRA. [25] It incorporated a thermal and electrical model of the generator to calculate the performance as a function of the thermal conditions across the couples and the associated degradation. Contrary to the heritage model, the physics-based model relies on testing/modeling of degradation at the couples/insulation level. [25] The following section provides a detailed description of the RTG degradation mechanisms to better understand the input needed for such a physics-based model.

9.3.1   RTG's degradation mechanisms

Mechanisms causing an RTG to lose power are summarized in Figure 9.7. As the fuel decays, the associated power drops about 0.8% per year. As a result of the fuel decay and the associated reduced heat input, the hot-junction temperature of the TE couples decreases (about 25°C over 17 years) resulting in a decrease in power output of the generator because of the lower temperature gradient across the thermoelectrics. The associated power drop from the RTG amounts to about 0.5% per year.

The third category includes the degradation mechanisms associated with the TE couples and insulation surrounding them. These degradation mechanisms include: 1) sublimation of the TE materials at the hot side of the couples, 2) change in TE properties (Seebeck coefficient, electrical resistivity, and thermal conductivity, 3) increase in electrical and thermal contact resistance at the couple interfaces, and 4) increase in thermal insulation conductance.

Sublimation of TE materials is more prominent on the hot-side of a TE couple because of high temperatures. Sublimation decreases as temperatures drop in the gradient from hot-side to cold-side temperatures. The sublimation rate is typically temperature-dependent and, if not controlled, could lead to a "necking" of the TE segment near the hot-junction that can cause an increase of the electrical resistance of the TE segments, which decreases the power output of the TE couples.

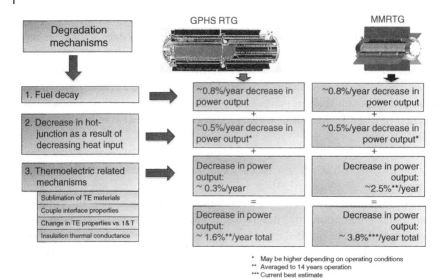

Figure 9.7 Radioisotope thermoelectric generator power degradation mechanisms. NASA / Wikimedia Commons / Public domain

Also, the sublimation products could travel and condense within the insulation surrounding the TE couples and on the cold-side of the TE couples. In severe cases, these condensed sublimation products can cause both thermal and electrical shorts that can also reduce the power output of the generator. The sublimation can also undermine mechanical integrity between TE materials and the metallization/hot-shoe interfaces. The sublimation rates are measured in $g\,cm^{-2}\,h^{-1}$ and should typically be kept below $10^{-6}\,g\,cm^{-2}\,h^{-1}$ to minimize the potential degradation noted earlier. Of course, this goal depends to some extent on the geometry, particularly the cross-section of the TE couples. A Si_3N_4 coating was developed for SiGe TE couples, and applied to the upper part of the SiGe TE segments to control the sublimation rate at an acceptable level. [25] With PbTe-based couples, sublimation rate control is achieved by filling the TE couple's housing with an inert cover gas.

The Seebeck coefficient, electrical resistivity, and thermal conductivity of TE materials used in the couples might change over time and with varying operating temperature. Those changes must be quantified. The changes could affect the conversion efficiency and reduce the power output of the generator. They could also affect the temperature gradient between the hot- and cold-junction of a TE couple. The temperature gradient could decrease if the thermal conductivity of the materials increases. The temperature gradient affects conversion efficiency and power output. Figure 9.8 shows the established variations of the Seebeck coefficient, electrical resistivity, thermal conductivity, and thermoelectric figure of merit (Z) for n-type SiGe used in the GPHS-RTGs. For n-type SiGe, substantial changes in

Seebeck coefficient and electrical resistivity are observed in the ~ 300–800 °C temperature range because of the phosphorus dopant precipitation. [26] Similar changes are observed for p-type SiGe. [26] Although substantial changes are observed for the Seebeck coefficient and electrical resistivity (driven by a change in

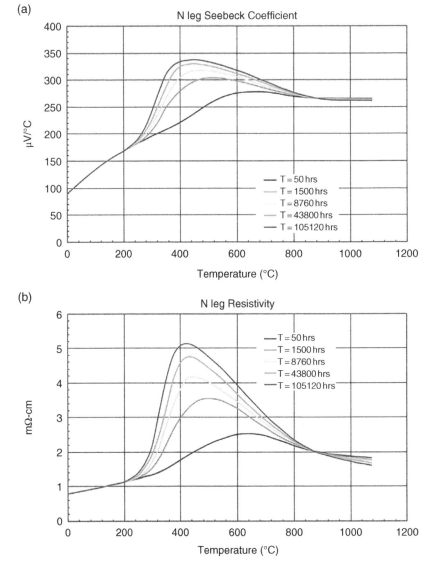

Figure 9.8 Variations of Seebeck coefficient (a) electrical resistivity, (b) thermal conductivity, (c) thermoelectric figure of merit Z, and (d) for n-type SiGe used in GPHS-RTG TE couples. [26]

(c)

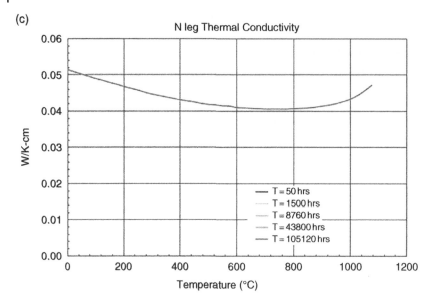

(d)

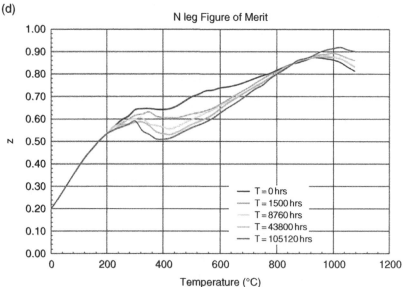

Figure 9.8 (Continued)

carrier concentration), the thermal conductivity, dominated by the lattice thermal conductivity, is not as sensitive to the changes in carrier concentration (dopant precipitation). Since the Seebeck coefficient and electrical resistivity vary in the opposite direction, the thermoelectric figure of merit ($Z = \alpha^2/\rho\lambda$) is fairly constant.

Detailed understanding of the variations in the Seebeck coefficient and electrical resistivity allows for an accurate prediction of the generator voltage and internal resistance. With MMRTG's thermoelectric couples, the TAGS materials segment undergo the majority of changes in its TE properties over the first ~ 10,000 hours.

The TE couples are composed of metallized n- and p-TE materials segments connected on the hot-side by a hot-side interconnect (sometime referred to as a hot-shoe when combined with a heat collector). The metallization stacks are typically composed of multiple layers, and each interface between these layers and the TE segments is potentially subject to mechanical/chemical degradation. This degradation could lead to an increase in the interface electrical contact resistance (ECR) and thermal contact resistance (TCR). An increase in ECR would increase the internal resistance of the couples, which would decrease the power output of the couples. An increase in TCR would decrease the temperature gradient across the TE segments and would also decrease the power output of the couples. ECR and TCR need to be characterized over time and temperature. This degradation mechanism is fairly small for SiGe TE couples but is a significant contributor to the degradation of PbTe-based couples.

Insulation materials surround the TE couples to maximize the flow of heat through the couples. The TE couples are operating at high temperature and a multifoil-based insulation was developed to minimize the radiation losses for the SiGe-based systems. A complex vapor transport degradation mechanism involving Si vapor from the hot-shoe and the quartz yarn used to wrap the SiGe couples was identified. [25] In this process, some of the sublimation products deposit on the cold-side of the couples. This transport of materials leads to a (thermal and electrical) shunt resistance forming at the cold-side of the couples, which decreases the power output of the couples. A bulk fibrous insulation is used for the PbTe-based generators as they operate at lower temperatures. The contribution of this degradation mechanism (i.e., change of thermal conductance of the bulk fibrous insulation as a function of time and temperature) to PbTe-based converters is not fully characterized and deserves more testing to fill that gap.

Besides these four main degradation mechanisms observed in heritage RTGs, other mechanisms could contribute to the degradation. Test campaigns aimed at providing the database needed for the physics-based life performance prediction model should include some initial testing (couples and coupons followed by DPA) to look for evidence of other degradation mechanisms beyond the typical degradation mechanisms previously described.

9.3.2 Physics-based RTG life performance prediction

A physics-based RTG performance prediction tool named DEGRA was developed to model the SiGe couple based RTGs. [25] The model has two major components. The first component is a computer code that contains an electrical and thermal model of the generator. The initial code was described by V. Raag. [27] To run the code, two sets of inputs are needed. The first set is the mission specific operating conditions, including the nominal GPHS heat source thermal inventory, the RTG fin root temperature (heat sink) as a function of time for the mission, and the RTG operating bus voltage. Based on this input, the thermal model in the code computes the hot- and cold-junction temperatures of the TE couples as a function of time. The second set of input needed, once the hot- and cold-junction temperatures are calculated, is the associated degradation at these temperatures. The code incorporates the changes in TE properties, ECR and TCR, sublimation, and insulation as a function of time and temperature and their impact on power output. The degradation database is critical for an accurate prediction. As previously mentioned, data needed for this database is typically generated from testing TE couples and coupons both at nominal and elevated temperatures. Testing at elevated temperatures is only valid as long as no new degradation mechanisms are introduced. RTGs can be used for multiple decades and testing of couples and coupons may only be performed for a few years before an RTG is launched on a mission. Yet, long-term predictions are needed to support mission planning. It is therefore critical to model the degradation mechanisms beyond the test duration and use the test data to anchor these models. Developing accurate physical/chemical models for degradation mechanisms is not always possible and semi-empirical models to reflect long-term degradation mechanism trends are sometimes needed as well. These models are incorporated into the RTG life performance prediction code.

A comprehensive test campaign using SiGe couples, 18-couples, and coupons [25] was performed to develop the initial degradation mechanisms database and models. Figure 9.9 shows the original DEGRA prediction for the Voyager

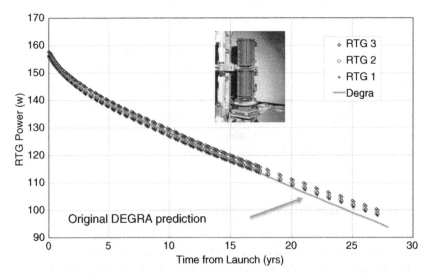

Figure 9.9 Comparison of Voyager RTG power output actuals and original DEGRA power prediction. NASA / Wikimedia Commons / Public domain.

RTGs, together with the RTG power output actuals. The agreement is excellent through ~15 years, after which the DEGRA prediction is slightly overestimating the degradation.

Today, the modern version of DEGRA JPL RTG physics-based code is called the Life Performance Prediction Model (LPPM). LPPM is a powerful tool for supporting new TE technology development and for mission planning and operations. Initially, the code was developed for SiGe-based RTGs. The LPPM code has since been updated to handle the MMRTG design and is used to support MSL and M2020 mission operations. It has been used to support development of new TE couple technology, including the skutterudite TE couples, which are the subject of Chapter 6 in this book.

9.4 Radioisotope Power System Dose Estimation Tool (RPS-DET)

The radioisotope power system dose estimation tool (RPS-DET) is a software application developed with collaboration between DOE, Oak Ridge National Laboratory, Idaho National Laboratory, and NASA's Radioisotope Power System Program to help researchers simulate and analyze ionizing radiation produced by an RPS. [28]

9.4.1 Motivation

The heart of every RPS is a radioactive heat source–or fuel. NASA/DOE-based RTGs use $^{238}PuO_2$ as a fuel, which generates heat through alpha-particle (α) decay. However, the radioactive emission of ^{238}Pu is not represented by α emissions alone. Additional particle emissions from this fuel are spontaneous fission photons and neutrons along with secondary reactions from α particles interacting with low-Z elements in the fuel to generate α-neutron (α-n) reactions. Many of these ancillary photon and neutron emissions are not attenuated by the fuel, escape the RPS, and are free to interact with materials, electronics, instruments, and personnel around an RPS. The ambient environment, fuel inventory and age, orientation of and proximity to an RPS all influence the effects of this emitted radiation. Certain scientific instruments, materials, operational considerations, research initiatives, design activities, and technology development programs affiliated with RPS may have radiological concerns that require an accurate and efficient method of estimating the effects of RPS-emitted ionizing radiation. RPS-DET was provided to researchers to expedite and simplify investigations of ionizing radiation emitted by an RPS.

9.4.2 RPS-DET Software Components

RPS-DET is a simulation and analysis application designed to provide researchers and designers not trained in radiation modeling, a platform for building, simulating, and analyzing three-dimensional Monte Carlo particle transport simulations of RPS radiation. RPS-DET is a MATLAB®-based graphical user interface (GUI) that serves as a "wrapper" for the SCALE 6.2.3 [29] nuclear software suite and also leverages various SCALE subsequences and utilities outlined in Table 9.2.

Two fundamental features of any RPS particle transport model are accurate physical geometries and time-dependent radiation source terms. RPS-DET provides users the ability to build and customize relevant RPS models based on pre-built geometry libraries, suitable default assumptions, and advanced customizations for isotopic and time-dependent fuel compositions.

Table 9.2 Brief description of various RPS-DET and SCALE software applications.

Software or Application	Brief Description	Use in RPS-DET	Reference
RPS-DET	MATLAB®-based graphical user interface	To build, execute, and analyze RPS-DET SCALE simulations	[28]
SCALE	Nuclear software suite developed at Oak Ridge National Laboratory	Provides industry-standard nuclear software for RPS-DET simulations	[29]
SCALE/ ORIGEN	Nuclear isotope decay software	Generates time-dependent neutron and photon source terms for PuO_2	[30]
SCALE/ MAVRIC	A coupled deterministic and Monte Carlo particle transport and shielding software	Performs neutron and photon particle transport calculations for RPS-DET PuO_2	[31]
KENO 3D	A geometry visualization tool for the SCALE software suite	Provides users the ability to render RPS-DET simulation geometries in 3D	[32]
Fulcrum	A graphical user interface to the SCALE software suite	Provides users an alternative method of editing input files, rendering geometries, and inspecting RPS-DET simulation results	[33]
Mesh View	A Monte Carlo mesh tally visualization and inspection tool	Provides users the ability to visualize and inspect 3D mesh tallies of RPS-DET radiation data with ease	[29]

9.4.3 RPS-DET Geometries

The geometry libraries for RPS-DET are subdivided into two categories: 1) RPS, and 2) environments. There are 29 assorted RPS geometries for the user to include in any of the 42 environments. A full list of all RPS-DET geometries is provided in Table 9.3.

Table 9.3 RPS-DET geometry library selections.

RPS Geometry #	RPS Geometry	Environment Geometry #	Environment Geometry
1–4	1–4 FCs[*]	1–2[†]	Concrete pad
5–22	1–18 GPHSs[**]	3	Earth's atmosphere
23	MMRTG[***]	4–9[††]	Concrete room
24	GPHS-RTG[+]	10	Inside the shipping container
25	ASRG[++]	11	Deep space
26–29	Modular Stirling[+++]	12	MSL-like aeroshell
		13	Cassini-like spacecraft
		14	New Horizons-like spacecraft
		15	Voyager-like spacecraft
		16–18[†††]	Moon
		19–25[†††]	Mars
		26–35[†††]	Titan
		36–42[†††]	Europa and/or Enceladus

RPS Notes:
[*] FC = fueled clad: represents the base-unit of every RPS-DET source term. Each iridium FC contains 151 g PuO_2.
[**] GPHSs = general purpose heat sources: aeroshell containers, each housing 4 FCs; modular and designed to accommodate vertical stacking inside RPSs.
[***] MMRTG = multi-mission radioisotope thermoelectric generator: currently available for NASA/DOE (2021).
[+] GPHS-RTG = general purpose heat source radioisotope thermoelectric generator: launched by NASA/DOE between 1989–2006. Certain flight systems still in spaceflight operation as of 2021.
[++] ASRG = advanced Stirling radioisotope generator: historic flight design of a cancelled Stirling convertor system.
[+++] Modular Stirling: conceptual Stirling RPS using vertical stack of GPHS in increments of 4 GPHSs (i.e., Models with 4, 8, 12, and 16 GPHSs).
Environment Notes:
[†] Directly on or suspended 1 m above a concrete pad (Earth atmosphere) (i.e., two options for RPS location in same geometry).
[††] Directly on or 1 m above a concrete floor, in the center, along the wall, or in the corner of a concrete room (Earth atmosphere) (i.e., six options for RPS location in same geometry).
[†††] Directly on, 1 m above, or buried/submerged in planetary regolith or ice (i.e., three options for RPS location in same geometry).

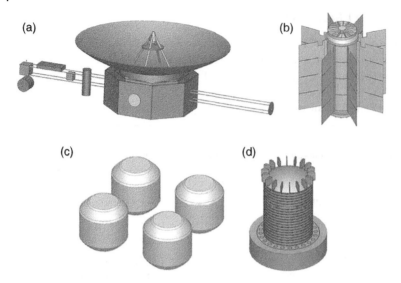

Figure 9.10 Examples of RPS-DET geometries. (a) Voyager-like environment, (b) MMRTG, (c) four fueled clads, and (d) terrestrial RPS shipping container.

The library of RPS-DET geometries is based on historic, current, and potential future RPS designs, coupled with relevant environmental conditions for the US space program. Any RPS geometry may be placed into any environment, and certain environments also allow for rotation of the RPS. This ability to combine geometric components provides a comprehensive simulation space for most traditional RPS-related scenarios, but RPS-DET also exposes the native SCALE input files for advanced users if custom alterations are desired. Renderings of various geometries are presented in Figure 9.10.

9.4.4 RPS-DET Source Terms and Radiation Transport

$^{238}PuO_2$ fuels consist primarily of ^{238}Pu, but other isotopes of Pu (i.e., ^{236}Pu, ^{239}Pu, ^{240}Pu, ^{241}Pu, and ^{242}Pu), their respective decay products, and secondary reactions in the fuel with various impurities (primarily α-n reactions) all influence the radiation field produced by an RPS's fuel at any point in time. Specifically, the spontaneous fission neutrons and photons, α-n reactions, and decay product photons all contribute to the overall external radiation environment with intensities and energy spectra also affected by the age of a specific fuel composition.

RPS-DET provides users with a reasonable default fuel composition (both for Pu isotopic composition and fuel impurities) based on published PuO_2 fuel specifications. [34] However, a user may opt to select and specify off-normal, or custom, fuel assays and impurity levels, if that level of detail is of interest.

Regarding time-dependance of the source term, $^{238}PuO_2$ has a complicated decay scheme. While spontaneous fission neutrons, photons, and α-n neutrons all follow a typical half-life decay corresponding to the 87.7-year half-life of ^{238}Pu, the daughter products of various Pu isotopes cause a buildup and energy shift of the emitted gamma spectrum for the first ~18 years of the fuel's life causing a multivariate and complicated source definition for modelers to assume.

To simplify this process, RPS-DET allows users to build simulations assessing RPS radiation either at a specific point in time (for dose-rate and flux calculations) or to time-integrate over a range of the fuel's life (for total integrated dose or fluence calculations). The ability to do this is enabled by SCALE/ORIGEN [30] generating unique fuel composition and time-specific source terms. By pre-populating default fuel assumptions and providing methods for defining more complex source terms, RPS-DET allows ease of generating instantaneous or multi-year source terms with only a few user selections. These source assumptions are then automatically imported into the Monte Carlo source term definition and transported accordingly using the SCALE/MAVRIC sequence. [31] While the default source term interaction simplifies the process for the novice radiation transport modeler, all native source input files (i.e., SCALE/ORIGEN and SCALE/MAVRIC) are made available as well and can be customized by advanced users.

9.4.5 Simulation Results

All RPS-DET simulations automatically include standard mesh tallies where superimposed uniform Cartesian grids composed of one million volumetric pixels–or *voxels*–overlay the entire simulation volume. Each voxel catalogs the number, energy, particle type, and path length of every particle that traverses the voxel through an entire simulation. These voxels (once automatically normalized to the source strength) are re-structured to create a 3D mosaic of the radiation field which can be viewed in 2D slices through various utilities leveraged by RPS-DET, Figure 9.11.

Each RPS-DET simulation produces multiple 3D mesh tallies representing neutron and photon flux (particles $cm^{-2} s^{-1}$). These flux tallies can also be combined with 26 separate response functions representing effective dose and other absorbed dose metrics in either English (e.g., rad and rem) or SI units (e.g., Gy and Sv). The user can easily toggle between these results, inspect data, and generate tables or line plots relevant to their individual analysis needs. Users can also add their own custom response functions to RPS-DET calculations.

(a) (b)

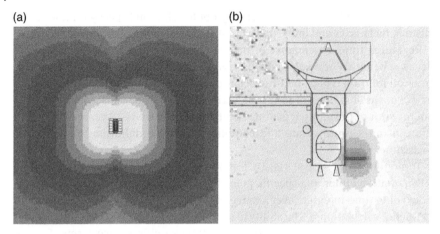

Figure 9.11 Examples of RPS-DET 2D cross-sectional slices of 3D mesh tallies, (a) and (b).

9.4.6 Validation and Verification

Verification processes were performed on >3,500 geometry combinations, source term assumptions, and object rotations to confirm compatibility with the entire simulation space provided by RPS-DET.

The accuracy of baseline assumptions for RPS-DET is based on the validation of multiple real-world, terrestrial radiation measurements of fueled flight units of the MMRTG. Modern RPS and fuel specifications were selected as validation cases because of a higher confidence in the accuracy and availability of detailed geometry information and up-to-date fuel assumptions. Default fuel compositions, fuel ages, and corresponding geometry selections agreed within 10% of measured values for the selected flight units.

9.4.7 Conclusion

The RPS-DET tool provides researchers with the ability to perform simulations of ionizing radiation from PuO_2-fueled RPS. RPS-DET leverages multi-agency input on geometries, fuel assumptions, radiation emissions, and environmental scenarios using the SCALE software suite, which adheres to an ASME NQA-1 quality assurance standard (i.e., a nuclear quality assurance standard). At the time of this publication, RPS-DET is available[6] to US citizens who have been approved by NASA's RPS Program Office and approved by the Radiation Safety Information Computational Center (RSICC). RPS-DET is maintained by developers at Oak Ridge National Laboratory (ORNL) and distributed by RSICC.

---

6 https://rps.nasa.gov/resources/rps_DET/

References

1 Woerner, D.F. (2017). Next-generation RTGs for NASA. *2017 AIAA Propulsion and Energy Forum, AIAA-2019-4612*, Atlanta, GA, (July 2017), 9 p. https://doi.org/10.2514/6.2017-4612

2 Ambrosi, R., Williams, H., Samara-Ratna, P., et al. (2013). Americium-241 radioisotope thermoelectric generator development for space applications. *2013 International Nuclear Atlantic Conference - INAC 2013*, Recife, Brazil (November 2013), 7 p.

3 Skrabek, E.A. and Trimmer, D.S. (1995). Properties of the general TAGS System. In: *CRC Handbook of Thermoelectrics* (ed. D.M. Rowe), 267–275. Boca Raton, FL: CRC Press.

4 Eggers, P.E. (1971). Advanced RTG and Thermoelectric Materials Study. *NASA-CR-122431*. Goddard Space Flight Center, Greenbelt, MD.

5 Fihelly, W. and Baxter, C.F. (1970). The SNAP-19 radioisotopic thermoelectric generator experiment. *IEEE: Trans. Geosci. Electron.* GE-8: 255–264.

6 Bennett, G.L. and Whitmore, C.W. (1989). On the development of the power sources for the Ulysses and Galileo Missions. *1989 Proceeding of the European Space Power Conference*, Madrid, Spain (October 1989), 117–121.

7 Stapfer, G., Rouklove, P., and Garvey, L. (1977). Progress Report No. 23 for a Program of Thermoelectric Generator Testing and RTG Degradation Mechanisms Evaluation. *DOE/ET/33003-T1*. Jet Propulsion Laboratory, California Institute of Technology, Pasadena, CA.

8 Stapfer, G. (1976). Degradation Model for an RTG with a Silicon-Germanium Thermopile. *DOE/ET/33003- T4*. Jet Propulsion Laboratory, California Institute of Technology, Pasadena, CA.

9 Stapfer, G. and Truscello, V.C. (1976). Long-term performance degradation of a radioisotope thermoelectric generator using silicon germanium. *Eleventh International Energy Conversion Engineering Conference (IECEC)*, State Line, NV (September), 1533–1538.

10 Kelly, C.E. (1975). HMW converter (RTG). *Tenth International Energy Conversion Engineering Conference (IECEC)*, Newark, DE (August 1975), 880–886.

11 Whiting, C.E. (2020). Proposed standard for analyzing and extrapolating in situ RTG and thermoelectric sub-assembly data to establish behavior trends and performance. *UDR-TR-2020-175*. University of Dayton Research Institute, Dayton, OH.

12 Whiting, C.E., Kramer, D.P., and Barklay, C.D. (2019). Empirical power prediction for MMRTG F1. *Proceeding of NETS 2019*, Richland, WA (February 2019), 51. American Nuclear Society.

13 Whiting, C.E. (2020). Empirical performance analysis of MMRTG power production and decay. *Proceedings under IEEE Conference on Aerospace 2020*, Big Sky, MT (March 2020), 2060. IEEE Publications.

14 Whiting, C.E. et al. (2022). *A Detailed Analysis of RTG Performance from 1961 to 2022*. University of Dayton: University of Dayton Research Institute.

15 Whiting, C.E. (2020). Understanding the MMRTG lifetime performance using modeling and analysis. *2020 APS4DS*. Pasadena, CA.

16 Bennett, G.L. and Lombardo, J.J. (1986). The general-purpose heat source radioisotope thermoelectric generator: power for the Galileo and Ulysses Missions. *21st International Energy Conversion Engineering Conference (IECEC)*, San Diego, CA.

17 Johnson, E.W. (1981). *GPHS RTG Program – RTG Temperatures*. Miamisburg, OH: Memo, Monsanto Research Corp.

18 Fountain, G.H., Kusnierkiewicz, D.Y., Hersman, C.B. et al. (2009). The New Horizons Spacecraft. In: *New Horizons* (ed. C.T. Russell), 23–47. New York, NY: Springer.

19 Anon (1998). Cassini RTG Program Final Technical Report. *DOE/SF/18852-T97*, Lockheed Martin Astronautics, Philadelphia, PA.

20 Bennett, G.L., Hemler, R.J., and Shock, A. (1994). Development and use of the Galileo and Ulysses power sources. *45th Congress of the International Astronautical Federation*, IAF-94- R.1.362. Jerusalem, Israel.

21 Whiting, C.E. Empirical Analysis of the Multi-Mission Radioisotope Thermoelectric Generator Qualification Unit Operated at Low Thermal Inventory with Potential for Improved End-of-Life Power. In: *Nuclear Technology*, 8 p. In Press.

22 Werner, J. and Otting, W. (2016). MMRTG User Guide. *INL/LTD-16-38850*. Idaho National Laboratory, Idaho Falls, ID.

23 Hammel, T., Bennett, R., Otting, W., and Fanale, S. (2009). Multi-Mission Radioisotope Thermoelectric Generator (MMRTG) and performance prediction model. *7th International Energy Conversion Engineering Conference*, Denver, CO (2–5 August 2009).

24 Woerner, D.F., Moreno, V., Jones, L., and Zimmerman, R. (2013). The Mars Science Laboratory (MSL) MMRTG in-flight: a power update. *Proceedings of Nuclear and Emerging Technologies for Space 2013*, Albuquerque, NM (25–28 February 2013).

25 Stapfer, G. (1976). Degradation Model for an RTG with a Silicon-Germanium Thermopile. *DOE/ET/33003-T4*. Jet Propulsion Laboratory, California Institute of Technology, Pasadena, CA.

26 Ekstrom, L. and Dismukes, J.P. (1966). Precipitation of phosphorus from solid solution in Ge-Si alloy. *Journal of Physics and Chemistry of Solids 27* (5): 857–863. https://doi.org/10.1016/0022-3697(66)90259-9.

27 Raag, V. (1971). Mathematical Model and Computer Program for the Design and Analysis of Silicon Germanium Air Vac RTGs. Memo #14. Syncal Corp.

28 Smith, M.B., Peplow, D.E., Lefebvre, R.A., and Wieselquist, W. (2019). Radioisotope Power System Dose Estimation Tool (RPS-DET) User Manual. *ORNL/TM-2019/1249*, 1560442. https://doi.org/10.2172/1560442.

29 Rearden, B.T. and Jessee, M.A. (2016). SCALE code system, ORNL/TM-2005/39, version 6.2.3. Oak Ridge Natl. Lab. Oak Ridge Tenn., Radiation Safety Information Computational Center as CCC-834.

30 Wieselquist, W. (2016). Capabilities of ORIGEN in SCALE 6.2.

31 Peplow, D.E. (2011). Monte Carlo shielding analysis capabilities with MAVRIC. *Nucl. Technol.* 174: 289–313.

32 Horwedel, J.E. and Bowman, S.M. (2000). *KENO3D Visualization Tool for KENO Va and KENO-VI Geometry Models*. TN (US): Oak Ridge National Lab.

33 Lefebrve, R. *Fulcrum User Interface*. Oak Ridge National Laboratory (ORNL).

34 Wong, S. (2001). Chemical analysis of plutonium-238 for space applications. *AIP Conference Proceedings* 552: 753–757.

10

Advanced US RTG Technologies in Development

Chadwick D. Barklay

University of Dayton Research Institute, Dayton, Ohio

10.1 Introduction

The invention of the radioisotope thermoelectric generator (RTG) is an enabling technology that allows spacecraft to venture to distant planets and beyond the boundaries of our solar system. Teams of researchers and scientists working at institutions worldwide direct a myriad of scientific instruments on these spacecraft and analyze the telemetry that flows back to unravel the mysteries of our universe. The discoveries associated with these missions would be impossible with more conventional power sources such as batteries, fuel cells, and photovoltaics. The successful performance of the RTGs on these missions laid the foundation for advanced power conversion technology development.

NASA's Radioisotope Power Systems (RPS) Program makes strategic investments in dynamic, thermoelectric, and other power conversion technologies to maintain NASA's current capabilities and develop advanced technologies. The latest NASA technology research roadmaps [1] include investments in space power and energy storage technologies, including thermoelectric materials. The development of conversion technologies in flying RTGs occurred in the mid to late 20th century. [2] NASA has made strategic investments to increase specific power to a range of 4 to 6 W_e/kg in RTGs yet that range appears to be an upper boundary [3] that remains unattainable however. The conversion technologies under development today promise to make RTGs more efficient, thus revolutionizing our ability to explore the solar system and interstellar space.

This chapter discusses the advanced thermoelectric technologies under development for space power systems. The development focus of these technologies is to improve long-term performance and manufacturability. Also discussed is

The Technology of Discovery: Radioisotope Thermoelectric Generators and Thermoelectric Technologies for Space Exploration, First Edition. Edited by David Friedrich Woerner.
© 2023 John Wiley & Sons, Inc. Published 2023 by John Wiley & Sons, Inc.

background on NASA's decision to resurrect the GPHS-RTG design and the generic technology challenges to reestablishing a SiGe unicouple production capability. This chapter concludes with a discussion of RTG concepts under development to support commercial space exploration.

10.1.1 Background

Technology innovation is critical for enabling ever more challenging space exploration missions. A broad range of factors drives technology development. These factors include strategy, policy, resources, performance goals, and technology push and mission pull. [4] Of these factors, technology "push" and mission "pull" require some explanation to contextualize their effect on the developing thermoelectric technologies. Mission "pull" is the demand for RPSs and focuses on meeting the requirements of planned or proposed missions. A technology "push" matures a specific technology for planned or proposed missions so as to deliver increased capability, reliability, or reduced cost. [5]

In the 1960s and '70s, RTG development efforts relied heavily on the "pull" for electrical power requirements of flight programs such as the Pioneer 10 and 11, Voyager 1 and 2, and Galileo missions. The "pull" from these missions drove development of the SNAP-19, Multi-Hundred Watt (MHW), and General Purpose Heat Source (GPHS) RTG systems. In addition, because of the requirements for increased electrical power for these missions, the missions's "pull" drove the development of thermoelectric materials and devices. For example, the SNAP-19 RTG employed lead-telluride to lead-telluride/TAGS (tellurium- antimony-germanium-silver) thermoelectric couples, and the MHW and GPHS-RTGs used silicon-germanium (SiGe) based thermoelectric couples.

However, in the late 1980s, NASA transitioned its research and technology development programs from responding to the "pull" from flight programs to the "push" of technology development to enable future missions. [5] As discussed, NASA's strategic investments in the "push" of thermoelectric conversion technologies have yet to yield a new flight-qualified high-performance RTG. The latest generation of flight-qualified RTG is the Multi-Mission RTG, which uses lead-telluride/TAGS thermoelectric materials developed for the SNAP-19 RTG. Over the past decade, NASA has become more responsive to the mission and science communities by transitioning their RPS technology development programs to a more nuanced version of technology "pull." One impetus behind this transition is the NASA Science Mission Directorate's Planetary Science Decadal Survey generated by the National Academies of Sciences, Engineering, and Medicine, which outlines a 10-year plan for scientific missions and goals. [6] This survey provides a framework for the NASA RPS Program to prioritize resources for developing and maturing RPS technologies to support notional missions documented in the study.

The Decadal Survey results in near-term "pull" for the RPS Program and mid-and long-term "pull" of technologies required in two to three decades. For example, the Next Generation RTG Project is pursuing a baseline technology that uses silicon-germanium thermoelectric materials, a near-term technology pull. As a mid-term technology "pull," the project plans to execute a development program to upgrade the thermoelectric materials for this system by employing lanthanum telluride materials.

10.2 Skutterudite-based Thermoelectric Converter Technology for a Potential MMRTG Retrofit

Authors:

Thierry Caillat[a], Stan Pinkowski[a], Ike C. Chi[a], Kevin L. Smith[a], Jong-Ah Paik[a], Brian Phan[a], Ying Song[b], Joe VanderVeer[b], Russell Bennett[b], Steve Keyser[b], Patrick E. Frye[c], Karl A. Wefers[c], Andrew M. Lane[c], and Tim Holgate[d]

[a] *Jet Propulsion Laboratory/California Institute of Technology, Pasadena, California*
[b] *Teledyne Energy Systems, Inc., Hunt Valley, Maryland*
[c] *Aerojet Rocketdyne, Canoga Park, California*
[d] *John Hopkins University Applied Physics Laboratory, Laurel, Maryland*

10.2.1 Introduction

One path to extending the current suite of available, flight-proven RTGs is a concept called the enhanced MMRTG. This was proposed by NASA's JPL in 2013. The design concept retrofits the flight-proven, TRL 9, MMRTG that uses PbTe/TAGS thermoelectric couples with higher-efficiency thermoelectric (TE) couples based on skutterudite (SKD) TE materials. The balance of the system is virtually unchanged. A multi-organization team composed of Teledyne Energy Systems, Inc. (TESI), Aerojet Rocketdyne (AR), and JPL is collaborating to develop and mature the SKD-based thermoelectric converter technology and perform the supporting systems engineering. The team's goal is to establish the potential for the eMMRTG concept to deliver a minimum of 77 W_e after 17 years (i.e., three years under storage conditions and fourteen years of flight). Several iterations of SKD couple development have been completed along with several years of systems engineering that updated radiation analyses and analytical models such as the dynamic, power, and thermal models of the generator. Each iteration produced a configuration that enhanced the robustness of the couple design and their strong potential to achieve the eMMRTG life performance goals. This section discusses the eMMRTG concept, reports on the most recent life performance test data for the couples and 48-couple module, and discusses the potential of the baselined SKD couples to meet the 77 W$_e$ power output goal at 17 years.

The flight-proven MMRTG comprises sixteen 48-couple modules composed of PbTe/TAGS thermoelectric couples packaged in fibrous insulation and operated under inert gas in a hermetically sealed environment containing O$_2$ and H$_2$ getters. The MMRTG uses the decay of PuO$_2$ fuel as thermal input for the 48-couple modules and produces about 118 W$_e$ at the beginning-of-life (BOL). The potential eMMRTG is nearly identical to the MMRTG except for substituting the upgraded 48-couple SKD modules for the PbTe/TAGS 48-couple modules. Figure 10.1 illustrates these and other potential changes.

One design change is to replace PbTe/TAGS-based couples with more efficient, higher temperature capable SKD-based couples. These SKD-based couples operate at a hot-junction temperature of 575°C, which is a 65°C increase compared to PbTe/TAGS-based couples. The second change is to replace a fibrous insulation used in the MMRTG thermoelectric modules with SiO$_2$-based aerogel. Modifying the isolation liner inner surface finish to maintain an acceptable temperature and structural integrity for critical system components is the third change.

One of the key performance requirements for the potential eMMRTG is its capacity to produce at least 77 W$_e$ of power output after 17 years of operation at the following conditions: T$_{fr}$ = 157°C, Q$_{BOL}$ = 1952 W$_{th}$, and V$_{load}$ = 34 V (T$_{fr}$ is the fin root temperature, Q$_{BOL}$ is the PuO$_2$ fuel inventory at BOL, and V$_{load}$ is the operating load voltage). The approach to verify this requirement relies on a combination of life test data gained from TE couples and devices and a physics-based life performance prediction model. Chapter 7 describes this model and testing approach in more detail.

10.2.2 Thermoelectric Couple and 48-Couple Module Design and Fabrication

Renewed interest in skutterudite (SKD) materials for thermoelectric power generation emerged in the early 1990s following JPL's pioneering investigations of several of these compounds. [7] Over the intervening decades, a wide range of scientific and application-related research and technology development efforts have occurred on these material compounds. SKD materials belong to a rather large family of compounds and solid solutions, whether filled or unfilled, offering many opportunities to tune their electronic and thermal properties. They are optimal for TE applications in the 300°C – 700°C temperature range with thermoelectric figure-of-merit peaks greater than unity. SKDs are a strong for a thermoelectric material. [8] Uher recently published a comprehensive review of skutterudite materials, which provides a wealth of fascinating properties of these materials. [9]

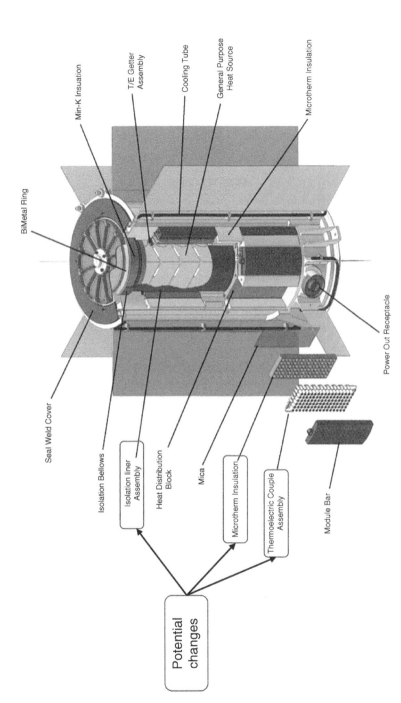

BiMetal Ring

Seal Weld Cover

Isolation Bellows

Min-K Insuation

T/E Getter Assembly

Cooling Tube

General Purpose Heat Source

Microtherm insulation

Isolation liner Assembly

Heat Distribution Block

Mica

Microtherm Insulation

Thermoelectric Couple Assembly

Module Bar

Power Out Receptacle

Potential changes

Figure 10.1 Changes under consideration for the potential eMMRTG. Credit: NASA/JPL–Caltech.

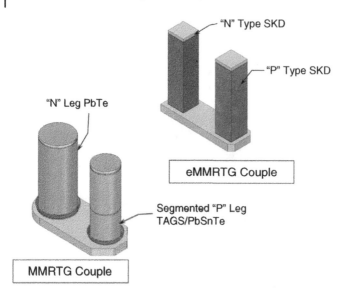

Figure 10.2 Illustration of the MMRTG and eMMRTG thermoelectric couples. Credit: NASA/JPL-Caltech.

The SKD materials for the eMMRTG concept are $CoSb_3$ and $Ce_{0.9}Fe_{3.5}Co_{0.5}Sb_{12}$ for the n- and p-type materials, respectively. Figure 10.2 illustrates the differences between the heritage MMRTG and eMMRTG couples. The length of the couple is identical to preserve the same spring-loading of the modules in the generator. However, because of manufacturing constraints, the cross-section of the SKD couples is square instead of circular, like the MMRTG couples. The p-leg of the SKD couple is segmented to optimize the mechanical robustness and TE performance. In addition, the design of the couple's cross-sections achieves an optimal hot-junction temperature for the thermoelectric that balances degradation over time and power output.

Over the last several years, the project team has conducted several development cycles on the couple, including fabrication and testing, to arrive at the current baseline design. The JPL and TESI team has demonstrated the manufacturability of these couples with a sufficient yield to support the potential production of an eMMRTG unit within the same duration that it takes to produce an MMRTG unit. TESI has fabricated over 200 baseline couples to support structural and life performance testing and requirement verification. Figure 10.3 shows a photograph of two baseline SKD couples. BOL performance testing of these couples demonstrates that their power output is within a few percent of the predicted values.

TESI has fabricated several Mini-Modules Life Testers (MMLT), incorporating 12 couples and also manufactured one 48-couple module. Figure 10.4 shows the

Figure 10.3 Photograph of the SKD couples currently on test at JPL and TESI. Credit: Teledyne Technologies Incorporated.

(a)

(b)

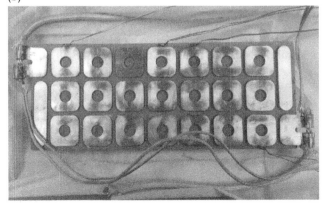

Figure 10.4 The first fabricated SKD 48-couple module. (a) is the top of the module; (b) is the underside of the module. Credit: Teledyne Technologies Incorporated.

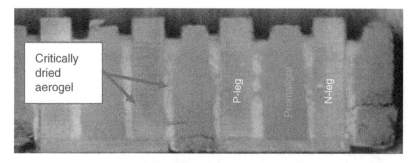

Figure 10.5 Cross-section of a 48-couple module showing the super-critically dried SiO_2-based aerogel (white areas around the legs) and the Promalight® bulk fibrous insulation (tan areas). Credit: Teledyne Technologies Incorporated.

first 48-couple module fabricated for testing. The couples are connected in series-parallel to augment reliability. Figure 10.4a shows the hot-shoes for all 48 couples; 10.4b shows the cold-side straps and the end terminals. A combination of super-critically dried SiO_2-based aerogel and bulk fibrous insulation (Promalight®) encapsulates the individual couples in the module. Figure 10.5 shows the critically dried aerogel, adjacent to the legs, is cast around the legs, establishing intimate contact with the couple legs. This level of encapsulation provides a sublimation barrier to keep the antimony sublimation to an acceptable level for long-term operation. The bulk fibrous insulation and SiO_2-based aerogel provide a low thermal conductance path around the couples that drives heat into the TE legs. As a result, the generator has a thermal efficiency of about 90%. The SKD 48-couple module represents the first development of an RTG thermoelectric converter in over 50 years since PbTe/TAGS RTGs were developed in the 1960s and SiGe-based RTGs in the 1970s!

10.2.3 Performance Testing of Couples and 48-Couple Module

Under the following conditions: $T_{fr} = 157°C$, $Q_{BOL} = 1952$ W, and $V_{load} = 34$ V, the hot-junction of the couple legs is at 575°C. With the radioisotope fuel decaying over time, the hot-junction temperature will decrease by approximately 30°C over 17 years. However, potential mission environments may cause higher hot-junction temperatures. Therefore, it is necessary to test above and below 575°C to characterize the couple's degradation rate as a function of time and temperature. Accelerated testing is an option when higher temperatures do not introduce new degradation mechanisms. Accelerated testing can contribute to establishing a 17-year performance prediction for the generator with more confidence.

JPL and TESI are testing SKD couples at 550, 575, 585, 600, 625, and 635°C using two MMLTs and several individual couples. The collected data includes open-circuit voltage (E_{oc}), internal resistance (R_i), and peak power for individual p- and n-legs and couples. Testing of the articles was started at different times and therefore the articles have different cumulative test durations with the longest over 20,000 hours.

For p-legs, the peak power, shown in Figure 10.6, correlates well with temperatures, and an increased level of degradation is apparent for higher temperatures. The peak power performance of the n-legs, shown in Figure 10.7, is more stable than the performance of the p-legs, at temperatures at or below 575°C, which controls the ability to meet the 17-year EODL requirement. Considering that the n-leg of the couple generates about two-thirds of the power, its long-term performance significantly contributes to the ability to meet the 17-year power requirement.

Figure 10.8 shows the normalized power for the couples, together with the normalized 48-couple module data and the original time-dependent prediction. The generator-level life-performance prediction model used couple-level data to establish degradation parameters. Figure 10.9 shows the normalized measured and predicted power output of the 48-couple module. The measured data is in excellent agreement with the prediction up to 6,500 hours of data. The 48-couple module is tested in a housing that is prototypic of an eMMRTG and has less

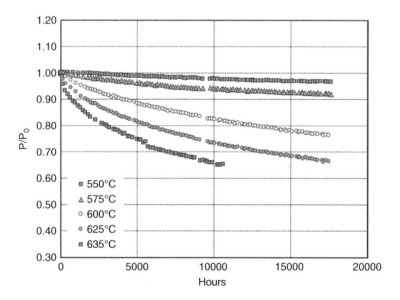

Figure 10.6 The normalized peak power of the p-legs as a function of time and temperature at 575 °C and below. Credit: NASA/JPL-Caltech.

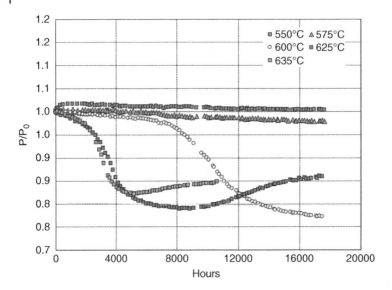

Figure 10.7 The normalized peak power of the n-legs as a function of time and temperature at 575°C and below. Credit: NASA/JPL-Caltech.

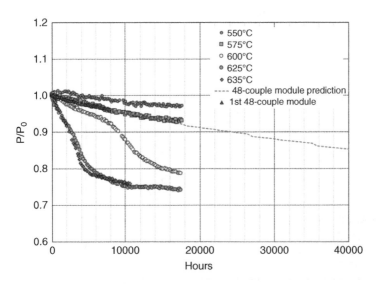

Figure 10.8 The normalized peak power of SKD couples as a function of time and temperature. The dashed line represents the 48-couple module data/prediction. Credit: Teledyne Energy Systems and the NASA/JPL-Caltech.

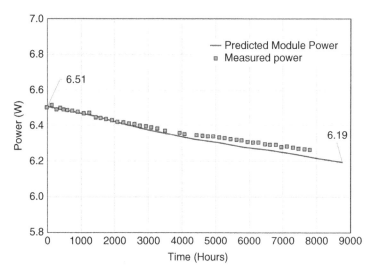

Figure 10.9 Measured and predicted power for the SKD 48-couple module is in excellent agreement. Credit: Teledyne Energy Systems.

instrumentation than the MMLTs. Therefore, the data from the 48-couple module is not used in the physics-based life performance prediction model, but serves as an initial verification of the life-performance prediction.

Additional testing of the couples for strength, thermal cycling, vibration, and irradiation resilience testing has also been completed. Details of the results are beyond the scope of this chapter, but the thermoelectric couples met the requirements. JPL is also planning to perform a vibration test on an aged 48-couple module.

10.2.4 Generator Life Performance Prediction

As discussed, the JPL and TESI team will verify the end-of-design-life (EODL) power requirement using couple life test data gained from thermoelectric couples/ devices and a physics-based life performance prediction model.

There are three fundamental mechanisms of an RTG's power degradation. The first mechanism is the decrease in hot junction temperature as the PuO_2 fuel decays, which results in a system-level degradation of about 0.8% per year; hot-junction temperature falls ~ 30°C over 17 years. This produces the second mechanism, a lower temperature gradient across the thermocouple legs, and results in a degradation rate of about 0.5%/year. The third mechanism is associated with the physical properties of the couples and their insulation, which includes 1) sublimation of the thermoelectric materials, typically at the hot-side of the couples, 2) changes in material properties (Seebeck coefficient, electrical resistivity, and thermal conductivity) over time and temperature, 3) increase in electrical and thermal contact resistance at the couple interfaces, and 4) increase in thermal insulation conductance. These potential degradation mechanisms are being quantified over time and temperature through testing, analysis, and modeling to enable long-term power output predictions.

The team has conducted up to two years of SKD TE property life testing, and the results show that the thermoelectric properties of the SKDs remain unchanged at temperatures ranging from 550 to 650° C. In addition, JPL has performed destructive physical analysis (DPA) of several SKD couples tested at temperatures ranging from 550 to 600°C. The DPA results show that the critical degradation mechanism of the p-legs, and to a much lesser extent of the n-legs, is associated with the couple hot-side interfaces between the SKD materials and the metallization layers. Degradation of these interfaces leads to an increase in the Electrical Contact Resistance (ECR) and the Thermal Contact Resistance (TCR) for the p-leg at these interfaces. The TCR correlates with reductions in open-circuit voltage observed for the p-legs over time, and the ECR correlates with the internal resistance increases for the p- and n-legs.

Therefore, it is necessary to quantify the ECR and TCR variations over time and temperature to best predict the degradation over time and predict eMMRTG power output. Predictions are based upon the ECR and TCR values collected during couple life tests and extrapolations are based on models derived from couple DPA results and test data. The DPA results are informative because they will show when the hot-side interface chemical reactions culminate and thus show when the degradation rates level off. As a result, JPL and TESI will refine the ECR and TCR models as more DPA and test data become available.

Figure 10.10 shows the current 17-year power output prediction for an eMMRTG and MMRTG based on the current ECR and TCR models that use test results from the n- and p-legs. The predictions assume three years of storage followed by fourteen years of operation and are a function of BOL fuel inventory (Q), generator fin root temperature (T_{fr}), and load voltage (V). The typical load voltage for the MSL and M2020 MMRTGs is about 30V. Under equivalent conditions, the eMMRTG could deliver at least 38% more power than the MMRTG. This significant improvement could enable more science for a spacecraft or rover using an eMMRTG rather than an MMRTG or could enable a spacecraft to carry, for example, two eMMRTGs instead of three MMRTGs, thus simplifying spacecraft integration tasks and lowering hardware costs.

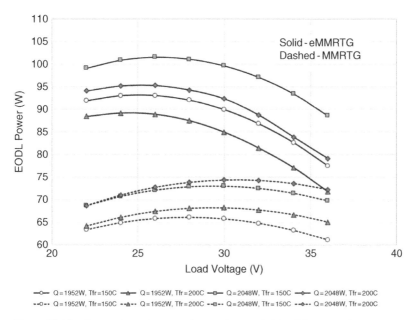

Figure 10.10 Comparison of current-best-estimate of eMMRTG and MMRTG 17-year power output as a function of beginning-of-life fuel inventory (Q), generator fin root temperature (T_{fr}), and load voltage (V). Credit: NASA/JPL-Caltech.

10.3 Next Generation RTG Technology Evolution

Chadwick D. Barklay

University of Dayton Research Institute, Dayton, Ohio

10.3.1 Introduction

As discussed earlier in this chapter, the design of the GPHS-RTG (Figure 10.11) represented the last significant technology evolution for a flight-qualified RTG. In the 1970s, the International Solar Polar Mission (ISPM), a joint mission between the European Space Agency (ESA) and NASA launched in 1990 as the Ulysses mission, provided a technology "pull" for the development of a more powerful RTG. [10] Building on the MHW-RTG technology, General Electric (later Lockheed Martin) designed and built the GPHS-RTG. Compared to the MHW-RTG, this new RTG design increased the output power from approximately 150 W_e to 300 W_e. [10] With a mass of about 55.9 kg, the GPHS-RTG's specific power is still the highest for a nuclear power system ever flown by NASA. [11]

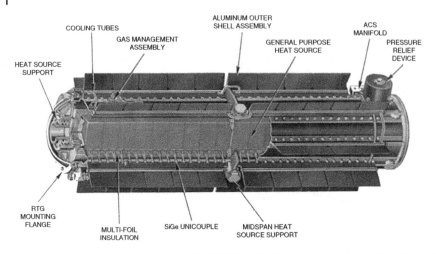

Figure 10.11 Cutaway diagram of a GPHS-RTG. Available from: https://en.wikipedia.org/wiki/GPHS-RTG

In 2014, the NASA RPS Program began the Nuclear Power Assessment Study (NPAS), which outlined a sustainable strategy to provide nuclear power systems for NASA's Science Mission Directorate (SMD) and Human Exploration and Operations Mission Directorate (HEOMD) missions for the next 20 years. [12] This study set the stage for a follow-up review of potential RTG concepts that could enable robotic scientific missions for planetary exploration over the next two decades.

In 2016, the NASA RPS Program initiated the Next Generation RTG Study, which JPL led. The study team included a diverse group of seventy-two scientists, engineers, technologists, and subject matter experts from NASA Glenn Research Center (GRC), NASA Goddard Space Flight Center (GSFC), JPL, Johns Hopkins University Applied Physics Laboratory (JHUAPL), and the US Department of Energy (DOE). The study's aim was to determine the RTG design characteristics that would "best" fulfill Planetary Science Division (PSD) mission needs. [4] The study examined a list of RTG capabilities and evaluated specific design attributes to identify candidate thermoelectric couple configurations and RTG concepts. The study used a bottom-up approach to identify eight high-efficiency thermoelectric couple configurations with lower technology development risk. Then, using the Decadal Survey as a guide, the study employed a top-down approach to identify three Next Generation RTG concepts that could best fulfill the potential power requirements of the conceptual missions outlined in the survey. [4] The study also examined requirements for specialized RTG concepts to support very challenging notional mission scenarios. For example, exploration of the ice sheets and likely oceans of the ocean worlds (such as Europa or Enceladus) or the surface of Venus would require a pressure vessel to protect the RTG from hydrostatic and dynamic pressure levels. [13] The results of the Next Generation RTG Study set the conditions for NASA to develop system and environmental requirements documents for the Next Generation RTG. The system requirements document defined the applicable design constraints, and the environmental requirements document captured the bounding environments for future mission destinations and launch vehicles. The release of the final documents allowed NASA to start a competitive procurement process for contractors to develop Next Generation RTG (NGRTG) concepts.

In early 2019, using a multi-phased acquisition approach, Idaho National Laboratory (INL) issued a competitive request for proposal (RFP) to develop the NGRTG. The initial effort required the contractor(s) to create a conceptual system design that included identifying the critical technologies necessary to fabricate the components and build the system. The follow-on phases of the NGRTG effort required fabrication of engineering, qualification, and flight units. Ultimately, for Phase 1 of the project, NASA selected Aerojet Rocketdyne and Lockheed Martin to develop NGRTG concepts.

However, as the design process for both contractors proceeded in mid-2020, the NASA SMD directed the NASA RPS Program to revector development of an NGRTG. SMD directed the use of SiGe unicouple technology and not other potentially, higher efficiency thermoelectric materials and configurations. The revectoring directive included reestablishing a "build to print" heritage SiGe materials and device production capability equivalent to those used for the GPHS-RTG program. Thus, the reference starting point for the NGRTG design became the heritage GPHS-RTG design but would use Step-2 GPHS heat sources. Since the Step-2 modules are thicker than the Step-0 and Step-1 modules used in the GPHS-RTG, this translates into a flyable thermal inventory of approximately 500 W_{th} less for an NGRTG Mod 1 design than the GPHS-RTG. The contractors reproposed and offered designs and plans to deliver a NGRTG Mod 1 RTG, and ultimately, in 2021, INL awarded a contract to Aerojet Rocketdyne to reestablish the GPHS-RTG production capability by 2027 (NGRTG Mod 1). In addition, eventually, Aerojet Rocketdyne will propose design modifications and enhancements to formulate the NGRTG Mod 2 RTG for implementation once the baseline or "build to print" NGRTG Mod 1 configuration achieves specific development milestones.

Concurrent with Aerojet Rocketdyne's NGRTG Mod 1 activities, NASA is also pursuing the feasibility of re-qualifying one of the defueled GPHS-RTGs for future flight operations and has dubbed that configuration NGRTG Mod 0. If Mod 0 proves flyable, then a fueled GPHS-RTG could be available to future missions earlier than 2027, and it would produce comparable power to the RTGs flown on NASA's Galileo and Cassini missions as it would house 18 GPHSs. Mod 1 design studies are ongoing. The unit may be limited to housing 16 of the larger and available Step-2 GPHSs because of cost or risk of elongating the housing to hold 18

Step-2 GPHSs. Consequently, the Mod 1 RTG power output currently is estimated over a range of ~230 W_e to 290 W_e at beginning-of-life (BOL) and up to 210 W_e at end-of-design-life (EODL), 17 years after BOL, including up to 3 years of ground storage. [5] The primary goal for the NGRTG Mod 0, Mod 1, and Mod 2 configurations is to provide missions with a reliable system with performance degradation behavior commensurate with GPHS-RTG. The Mod 2 configuration targets 290 W_e at EODL, assuming a performance degradation comparable to Mod 0 and Mod 1, translating to 400 W_e at BOL. [14]

10.3.2 Challenges to Reestablishing a Production Capability

The Ford Model T was one of the first cars to be produced via an assembly line, and there were 84 individual steps required to assemble approximately 1,500 parts into a car. In 1913, the company trained factory workers to specialize in one of the assembly steps. Today, the average car has roughly 30,000 parts, and automobile assembly lines are more automated and complex. However, imagine if a major automobile manufacturer other than Ford decided to mass-produce a Model T today. Consider all the challenges to establishing the assembly process, designing and building tooling, setting up supply chains, training assembly workers, and programming computers to replicate a process that culminated almost one hundred years ago. That analogy is not to infer that the GPHS-RTG is the modern-day equivalent of a Ford Model T. Instead, it highlights that Aerojet Rocketdyne's challenges in establishing a "build to print" GPHS-RTG production capability are not insignificant.

10.3.2.1 Design Trades

One of the most significant design-related decisions that both Aerojet Rocketdyne and Lockheed Martin encountered was responding to the revectoring directive. Compliance with the directive included reestablishing a factory or factories to "build to print" the heritage SiGe materials and device production capability equivalent to the Cassini-era capability employed by Lockheed Martin in the 1990s. As a result, both competing contractors assessed a design approach that they believed best fulfilled the revectoring directive. Although it is impossible to discern the processes that either contractor underwent during this process, it is logical to assume they evaluated specific design trades between Step-2 modules and output power. A design trade is when one system attribute is identified as less significant than another; thus, the more critical attribute has more influence on a design process. Here, each contractor resolved whether "build to print" meant maintaining the geometric proportions of the GPHS-RTG and using 16 Step-2 GPHS Modules or elongating the GPHS-RTG design to accommodate 18 Step-2 GPHS Modules and incorporating more thermoelectric unicouples into the design.

As discussed, the use of 16 Step-2 modules translates into a thermal inventory of approximately 500 W_{th} less for the NGRTG Mod 1 design than the GPHS-RTG, which translates into a reduced BOM power output of roughly 20-30 W_e. If the design trade supported elongating the housing to accommodate 18 Step-2 modules, then recovery of 20–30 W_e of BOM power output is feasible. However, this trade would involve engineering changes beyond the intent of "build to print" and could adversely affect downstream operations during the fueling process at Idaho National Laboratory because of changes in tooling and hardware issues. Although mitigation of the potential downstream problems is easy to perform, the mitigation actions have a cost that transcends the value of the recovery of 20-30 W_e at BOM.

10.3.2.2 Silicon Germanium Unicouple Production

The GPHS-RTG has over 3,100 individual components and subassemblies, but resurrecting the fabrication processes for some of these components will be easier than for others. The greatest challenge to reestablishing delivery of the GPHS-RTG Mod 1 is production of the GPHS-RTG unicouple (Figure 10.12). There are 18 different fabrication and assembly steps in the GPHS unicouple assembly process.

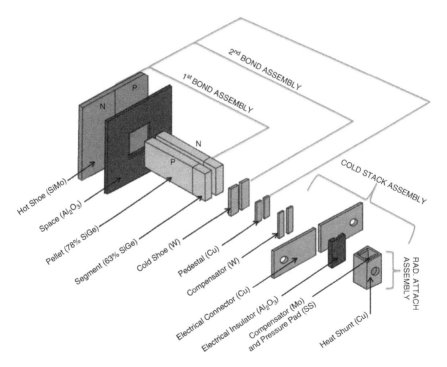

Figure 10.12 SiGe Unicouple Components. Available from: https://en.wikipedia.org/wiki/Application_of_silicon-germanium_thermoelectrics_in_space_exploration

During the Cassini program in the 1990s, unicouple production yields were consistent with production efforts during the Galileo and Ulysses programs in the previous decade. The production yields of each SiGe unicouple fabrication and assembly step ranged from 75.1% to 98.6%, resulting in an overall production yield nearing 50%. [15] It is essential to understand that the fabrication and assembly processes were well established in the 1970s and performed by highly trained personnel, yet resulted in a low yield. Those low yield rates further stress Aerojet Rocketdyne's challenges during the reestablishment of a "build to print" heritage SiGe materials and device production capability.

To mitigate the future challenges that a contractor will encounter during Phase 2 of the NGRTG effort, in mid-2020, the NASA RPS Program created a task force to assess risks of reestablishing a "build to print" heritage SiGe unicouple production capability. The task force included researchers and subject matter experts from the Johns-Hopkins Applied Physics Laboratory (APL), NASA Jet Propulsion Laboratory (JPL), Oak Ridge National Laboratory (ORNL), and University of Dayton Research Institute (UDRI). The task force reviewed various aspects of the SiGe unicouple fabrication and assembly processes and provided a qualitative risk assessment that included mitigation strategies and activities. Based on these initial efforts, the NASA RPS Program directed the task force to pursue specific risk mitigation activities identified by the task force. These risk mitigation activities included pathfinding efforts to fabricate SiGe materials, explore alternative fabrication technologies, and assemble SiGe unicouples from new and heritage materials and components.

Besides fabricating SiGe and SiMo components, the task force also performed casting, crushing, pulverizing, and blending SiGe materials. In addition, a significant level of effort focused on doping vacuum cast materials and understanding the effects of particle size, morphology, and distribution of powder blends on the microstructure of hot-pressed materials. The task force activities culminated at the beginning of Government Fiscal Year (GFY) 2022. Aerojet Rocketdyne incorporated some task force activities into their risk mitigation plan to reestablish the heritage SiGe Unicouple production capability. These activities include continuing some development activities and technology transfer of knowledge and lessons to transition the technology from a laboratory setting to production. Aerojet Rocketdyne will develop detailed specifications, procedures, and drawings from heritage material fabrication processes to establish controlled processes for that production of flight hardware.

10.3.2.3 Converter Assembly

The GPHS-RTG converter assembly comprises the thermopile and the converter shell. A thermopile converts some of the decay heat of radioisotope heat sources into electrical power. An outer shell provides containment and structural support for the heat source and thermopile, rejecting excess heat through radiating fins.

A GPHS-RTG thermopile consists of multifoil insulation, a molybdenum inner frame, two internal frame supports, 572 thermoelectric SiGe unicouples, and cold-side electrical straps. [15] A multifoil insulation is wrapped throughout the thermopile and surrounds the heat source. The insulation consists of 60 alternating layers of molybdenum foil and Astroquartz cloth. In addition, a thick metallic foil layer provides additional support on the inside of each end cap, and an inner molybdenum frame supports the axial section of the insulation. [15] The insulation assembly has penetrations for the unicouples and the inner molybdenum frame supports. The end caps also have penetrations for the heat source support preload studs.

SiGe unicouples are installed one at a time through the inside of the thermopile insulation. Then copper straps are bonded to each leg and riveted together to create an electrical circuit. The unicouples are each fastened to the converter shell or housing after installing the thermopile assembly into the housing. [16] Figure 10.13 is a photograph of the thermopile assembly awaiting insertion into a converter shell or housing.

After installing a thermopile into the converter housing, assembly of the outboard heat source support frame occurs, and then an Electric Heat Source (EHS) is inserted into the ETG. After EHS insertion, installing the inboard heat source assemblies follows, preloading the EHS. Next, leak testing of the ETG occurs after

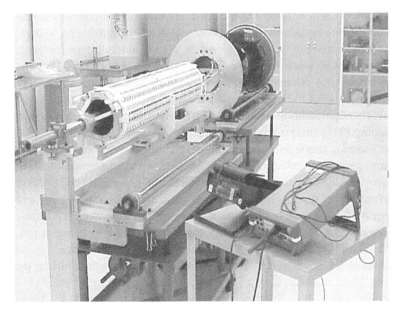

Figure 10.13 GPHS-RTG Converter Shell and Thermopile Mating Operations. [16] Harmon et al. (2007), AIP Publishing LLC.

installing the gas processing dome. [16] The converter assembly then undergoes a vacuum bake out to remove volatiles and is tested for electric performance.

Aerojet Rocketdyne will need to develop NGRTG-specific fixtures and assembly procedures based on heritage Lockheed Martin procedures and documents used during the Cassini and New Horizons. Although the converter assembly process described above appears straightforward, several assembly operations can cause increases in the resistance of the thermoelectric circuit that could cause the thermopile to exhibit unstable behavior. Misalignments in the multifoil insulation coupled with geometric variations in the penetrations for the unicouples and internal frame supports can lead to additional thermal losses. Even though Aerojet Rocketdyne has considerable experience in assembling the MMRTG converter, the architectural complexities of the NGRTG will challenge engineers during the fabrication of the Engineering Unit (EU).

10.3.3 Opportunities for Enhancements

As discussed earlier in this section, the revectored NGRTG effort allowed the contractors to propose design modifications and enhancements (NGRTG Mod 2) for implementation once the baseline "build to print" NGRTG configuration (NGRTG, Mod 1) achieved specific milestones. Potential improvements to the NGRTG Mod 1 can provide increased performance and greater efficiency. The Next Generation RTG Study published in 2017 by JPL included an evaluation of potential candidate NGRTG thermoelectric materials, some of which may be suitable for NGRTG Mod 2 enhancements. The study considered 38 n-type and 29 p-type materials whose properties were available in open-source scientific publications. [13] The technology maturity of the evaluated thermoelectric materials ranged from fundamental research level to flight-proven. The study found eight thermoelectric couple (TEC) configurations that have minimal risk and enough efficiency to warrant developing RTG concepts around them. [13]

However, several thermoelectric couple-level configurations evaluated had two- or three-segment architectures. [13] In addition, for most segmented architectures, there was little test data on materials compatibility issues related to direct bonding of segmented thermoelectric materials or on materials compatibility and thermal expansion mismatch issues related to bonding layers between segmented thermoelectric materials. These technology gaps lead to tradeoffs between programmatic risk and the desire for high potential system-level efficiencies.

Researchers at JPL have been investigating refractory rare-earth tellurides for several years. These materials have a Th_3P_4 structure and type have attracted considerable interest as high-performance thermoelectric materials since the 1980s because of their high dimensionless figure of merit (ZT). [17] JPL's lanthanum telluride ($La_{3-x}Te_4$) research shows that peak ZT values greater than 1.1 at 1,273 K are viable. [17] For this material, single-segment architectures are low risk and offer incremental efficiency improvements compared with MMRTG and predicted eMMRTG performance levels. However, two- and three-segment architectures have the potential for high efficiencies by comparison, but segmented architectures that use $La_{3-x}Te_4$ as the high-temperature segment present a spectrum of technology risks associated with bonding the individual thermoelectric material segments, depending on the configuration.

Segmented configurations will have to show thermal, structural, and chemical stability and maintain nominal performance levels during vibration and thermal cycling conditions. These characteristics are especially critical if NGRTG Mod 2 enhancements include segmented thermoelectric architectures in a cantilevered couple design. Unfortunately, these segmented material combinations' long-term stability and performance are unknown, especially under vacuum conditions. Sophisticated bonding and metallization layers could be required to join the segment interfaces and the segmented thermocouple legs to the cold- and hot-shoe materials. Sublimation suppression coatings may be essential to ensure the long-term stability of NGRTG Mod 2 thermoelectric couples. Coatings for this purpose are not unusual since the SiGe GPHS unicouples employed a silicon nitride coating.

If Aerojet Rocketdyne pursues NGRTG Mod 2 enhancements, a significant level of research, development, and testing will be required. There is an unquantifiable level of uncertainty associated with the level of technology development needed to mature potential NGRTG Mod 2 thermoelectric materials and architectures from their current TRL to a TRL greater than 2. The Next Gen RTG Mod 2 upgrade represents an evolutionary step in performance, but no plug-and-play options are available today.

10.4 Considerations for Emerging Commercial RTG Concepts

Chadwick D. Barklay

University of Dayton Research Institute, Dayton, Ohio

10.4.1 Introduction

In 2020, humans accessed space in a vehicle built and owned by a private corporation, SpaceX. This flight opened a new chapter of spaceflight led by private firms. [18] SpaceX, Blue Origin, and Virgin Galactic have launched people into space, but their long-term space aspirations transcend tourism by aiming at settlements.

There will be a demand for autonomous satellites, probes, spacecraft, rovers, and telecommunications stations as the commercial space economy develops and expands. Regardless, if future pioneers establish colonies on the Moon or Mars, various power systems will be required. These power systems will most likely combine solar, fission, and radioisotope technologies. However, on the Moon, solar power is not an option during the two-week lunar night and a serious challenge while operating in permanently shadowed regions, which makes fission and RTGs the only viable power sources for those environments. RTGs will play a prominent role in powering systems there.

NASA is also working with several commercial companies to deliver science and technology to the lunar surface through the Commercial Lunar Payload Services (CLPS) initiative. [19] These companies are designing and building spacecraft to deliver payloads to the Moon for NASA. The payloads could be bespoke surface systems, science investigations, and technology experiments, which will require reliable power technologies to survive the lunar nights. However, the MMRTG is the only flight-qualified RTG in the United States. NASA levies a cost on mission proposers to use one or more MMRTGs. In 2019, the mission cost for a single MMRTG was $54 million and an additional $15 million for a second unit, which does not include the cost of the National Environmental Policy Act (NEPA) and launch approval process support. [20] It is important to note that the mission costs for using an MMRTG do not recover the total cost. The infrastructure costs to produce and process ^{238}Pu for NASA are about $83 million per year. [24] These potential costs make an MMRTG not a practical option for a commercial entity. This factor, coupled with several others discussed further in this section, opens the trade space for commercial space RTGs.

10.4.2 Challenges for Commercial Space RTGs

NASA is the global expert in developing, qualifying, and flying RTGs for space exploration. A few other entities are also developing RTGs for space. Since 2009 the University of Leicester in the United Kingdom (UK) has been developing an americium-241 based RTG (^{241}Am RTG) for the European Space Agency. [21] Also, the Indian Space Research Organization (ISRO) and the Korea Atomic Energy Research Institute (KAERI) are each developing RTG capabilities for space exploration. ISRO plans to design and build a 100 W_e RTG but has not selected a specific radioisotope heat source material. KAERI is working with the University of Leicester and the UK National Nuclear Laboratory (NNL) to co-develop a milli-watt RTG that will employ a strontium-90 based heat source. [22] Commercial startup Zeno Power Systems is also developing strontium-90 based heat sources for terrestrial and space RTG systems (^{90}Sr RTG). Zeno Power Systems targets initial RTG power outputs in the 10 W_e range but intends to pursue higher power RTG designs. [23] The challenges for designers of these systems are obtaining the quantities of radioisotopes required for the design, maximizing the specific power, and navigating the nuclear launch approval process.

10.4.2.1 Radioisotopes

NASA uses plutonium-238 as a heat source material for space RTGs. However, this material is in short supply, expensive to produce and process, and all facets of the DOE infrastructure supply chain are tailored for this radioisotope. Plutonium-238 is made in a nuclear reactor, whereas strontium-90 and americium-241 are extracted from spent (used) nuclear fuel. The difference in production processes for these radioisotopes makes strontium-90 and americium-241 less expensive than plutonium-238. The United States does not recycle spent fuel and the DOE cannot produce sufficient quantities of americium-241 to fuel even the smallest of space RTGs. It is difficult for the DOE to process older, low assay strontium-90 in amounts required for space RTGs with more than a few watts of electrical power output.

However, France, the United Kingdom, Russia, India, and Japan reprocess spent fuel. Thus, strontium-90 and americium-241 are viable radioisotopes for developing space RTGs for both government and commercial entities. Plutonium-238 has a half-life of 87.7 years, making it an ideal radioisotope for RTGs used for deep space exploration. Strontium-90 is not a suitable radioisotope for RTGs used for this purpose. Its half-life is 28.8 years; thus, a significant amount of its thermal inventory decays in less time than it takes to complete a typical NASA deep space mission. However, strontium-90 could be a viable radioisotope for a space RTG if the mission duration is three to seven years in length. On the other end of the spectrum, americium-241 is a suitable radioisotope for RTGs used for deep space exploration because of its long half-life of 432.2 years.

10.4.2.2 Specific Power

Of interest to mission planners is the specific power of the RTGs powering their spacecraft. For an RTG, specific power is ratio of electrical power output and the mass of the unit (W_e/kg). As a rule of thumb, for a NASA space mission, the payload launched (satellites, spacecraft, etc.) is about 6% of the total mass of the launch, and the remaining 94% of the mass is the launch vehicle and propellants. The payload mass can dictate the selected launch vehicle and potential launch windows, depending on the mission. Thus, in an ideal situation, mission planners prefer to use a power system with the highest available specific power.

Employing higher power thermoelectric devices, radioisotopes with a higher thermal power density (W_{th}/cm^3), and reducing the mass of the other components of an RTG can increase the specific power of an RTG. For perspective, the specific power of the GPHS-RTG and the MMRTG are 5 W_e/kg and 2.5 W_e/kg respectively. [25] The goal for the Next Gen RTG Mod 1 system is to achieve the same specific power as the GPHS-RTG, designed in the 1970s. For the NGRTG Mod 2, the only way to increase specific power is to upgrade the thermoelectric conversion technology with a higher power output than silicon germanium. The ^{241}Am RTG designers are targeting a specific power of about 1.0 W_e/kg [18]. Because of the lower specific power of this type of RTG, it is better suited for systems with lower electrical power outputs of 10–25 W_e. Increasing the specific power of a ^{241}Am RTG is achievable by minimizing thermal losses, reducing the mass of the structural components of the converter assembly, and increasing the efficiency of the thermoelectric conversion technology. However, any improvements to the specific power of this type of RTG are incremental. There are more significant opportunities for specific power gains in ^{90}Sr RTGs designed for space. A reasonable specific power target for these systems is 0.25 W_e/kg, which is an order of magnitude greater than legacy, ^{90}Sr terrestrial systems. Similar to other systems, higher efficiency thermoelectric devices and reductions in mass are also viable strategies for increasing specific power. However, using high-assay strontium-90 (~50% ^{90}Sr) will increase the specific power by reducing the volume of the encapsulated radioisotope and enable the shrinking of the size of an RTG, which reduces the RTG mass by minimizing the volume of shielding required. Another method of increasing the specific power of ^{90}Sr RTGs is to minimize the bremsstrahlung radiation generated within the fuel [26] so that the mass of the shielding required to protect the personnel assembling, testing, handling, and transporting the RTG can be lowered.

10.4.2.3 Launch Approval

In 2019, President Trump issued NSPM-20, "Presidential Memorandum on Launch of Spacecraft Containing Space Nuclear Systems," which updated the federal process for launch of space nuclear systems by establishing risk-based safety analysis and launch authorization processes. [27] NSPM-20 defines a 3-Tier set of categories based on radioactive material-at-risk, technology, and radiological risk estimates stemming from a nuclear safety analysis. [28] However, after completing a nuclear safety analysis, the determination of the final tier level determination is made. The A2 value of the radioactive material, which is a value used in transportation regulations as a normalized measurement of radiological risk, is the guideline for determining the amount of material-at-risk. [29] Figure 10.14 illustrates the NSPM-20 tier criteria; the A2 values are across the top of the figure. For context, a NASA launch of the MMRTG is a Tier III launch, although the A2 value is in the Tier II category. It becomes a Tier III launch because the safety analysis indicates a credible accident scenario (≥ 1/1,000,000 probability) that results in radiation exposure greater than 25 rem total effective dose to any member of the public.

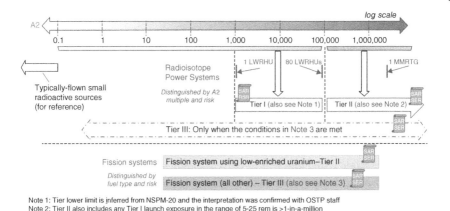

Figure 10.14 Illustration Showing NSPM-20 Tier Factors. [28]

NSPM-20 established that the Department of Transportation (DOT) has statutory authority to license commercial space launches that use Space Nuclear Systems. The DOT's Federal Aviation Administration (FAA) licenses all commercial space launches and reentries and the operation of commercial launch and re-entry sites. This delegation of authority for commercial launches that employ radioactive materials streamlines the launch approval process and makes it consistent with non-nuclear launches. However, this delegation of authority presents potential challenges for commercial RTG providers.

First, the FAA does not have the internal subject matter expertise to review safety analyses and issue a safety evaluation report (SER). As a result, the FAA employs the Interagency Nuclear Safety Review Board (INSRB) to perform those activities. Therefore, one of the initial challenges for potential RTG providers is the initial engagement with the FAA and INSRB after the mission's conceptual design. In the past, these initial meetings regarding non-nuclear launches only included the launch service and payload providers. Thus, integrating RTG providers and INSRB into this forum is critical to identifying SER timing issues for mission planners to address without creating unnecessary delays in a launch schedule.

The next challenge for commercial RTG providers is navigating the 3-Tier launch approval approach. Before the publication of NSPM-20, the launch approval process was a one-size-fits-all approach equivalent to the current Tier III category. The MMRTG, the only flight-qualified RTG, employs a well-characterized plutonium-238 encapsulation and aeroshell safety system. However, despite this high level of characterization and safety testing, the launch approval process for the MMRTG still takes several years. [28] The RTG launch approval process for NASA launches takes an average of 6.5 years and costs an estimated $40 million, including

$8.2 million for radiological contingency planning. [30] The cost and schedule length for this activity could prohibit a commercial mission flying an RTG or other Space Nuclear System (SNS). Commercial RTG providers will likely utilize radioisotopes other than plutonium-238 encapsulated in aeroshell safety systems not flown before. Thus, it will be a challenge to establish what level of safety analysis constitutes "good enough" to satisfy the intent of Tier I and Tier II safety analysis.

10.4.3 Launch Safety Analyses and Testing

The risks associated with launch or re-entry accident scenarios are often analyzed and modeled using computer simulations informed by impact testing during the launch approval process for RTG systems. These analyses form the foundation of the mission-specific safety analysis report (SAR), which documents the potential risk and adverse effects of radiological dispersion over populated areas. [29] The SAR is a critical document that provides decision-makers with the tools and information necessary to approve the launch.

10.4.3.1 Modeling Approaches

The safety analysis modeling approach can involve deterministic, probabilistic, and stochastic models. A deterministic analysis is an empirical approach that uses conservative assumptions and calculations to assess the adequacy and performance of safety systems. [31] A probabilistic safety analysis considers a comprehensive list of potential accident scenarios and selects those exceeding specific criteria to identify and assess the adequacy of safety systems. [31] NASA began using probabilistic risk assessment methods in 1967 after the Apollo 1 accident. [31] The safety analyses modeling for the Curiosity and Perseverance missions employed stochastic models with Monte Carlo simulations to provide probabilistic results of the potential health risks resulting from the use of the MMRTG. [32] However, as discussed in this chapter, this type of safety analysis is prohibitively expensive and takes several years to complete. As a result, DOE and NASA have embraced a more deterministic approach for evaluating the launch safety of RTG systems. This change in direction is workable because the GPHS Module (see chapter 7) has over forty years of safety testing, ranging from impact testing of individual fueled clads to a complete RTG system.

However, for new commercial RTG systems that possess very little safety testing data, the use of a cost-effective deterministic modeling approach is not viable. Using a purely probabilistic approach is not feasible because of the cost and duration of the analyses. Thus, the challenge for commercial RTG providers is to devise a hybrid process that employs deterministic techniques to identify worst-case accident scenarios and a probabilistic-lite modeling approach to determine the potential health risks resulting from the use of the commercial RTG system.

10.4.3.2 Safety Testing

A commercial RTG's radioisotope encapsulation will have to maintain its integrity during various accident scenarios with accident conditions including blast (over-pressure and fragmentation), propellant fire, and post-atmospheric re-entry impact events. Impact testing of an RTG's components simulates the environment, temperature, and mechanical insult severity for specific accident scenarios. In the past and within current practice, safety testing of GPHS heat source components uses encapsulated fuel with radioactive and non-radioactive simulants.

Some impact tests subjected an entire RTG, fueled with non-radioactive simulant materials, to simulated rocket booster fragment impacts using a rocket sled [33]. In contrast, other booster fragmentation tests involved plates impacting a GPHS Module. However, most impact tests were conducted on the radioisotope encapsulation boundary, the metal clad around a fuel pellet, aka, a bare clad. The aim of these tests was to assess the integrity of the encapsulation boundary to various overpressure and impact conditions (temperatures, velocities, and orientations). If the encapsulation boundaries ruptured, post-mortem examinations analyzed the size, morphology, and distribution of fuel particles to understand the level of contamination in the simulated accident scenario.

For commercial RTG providers, conducting the level of impact testing and analysis comparable to the GPHS-RTG program will be challenging, if not impossible. This problem is especially true if impact testing employs actual radioisotopes and not non-radioactive surrogate materials. Further, the intent of many of the bare-clad impact tests was to improve the knowledge-base and better compare the response of the radioisotope and its surrogates. Thus, finding a suitable non-radioactive surrogate material for impact testing is paramount to ensure that the response of impact testing on the encapsulation boundary is consistent with the response of a proposed radioisotope. For a ^{90}Sr RTG, this issue becomes extraneous because the stable form of strontium-90 is strontium-88. Both materials chemically behave the same; thus, all impact testing could employ the non-radioactive strontium-88. Americium is a synthetic radioactive element, and americium-241 does not have a non-radioactive, stable sister as does strontium-90. However, the team at the University Leicester is using encapsulated neodymium sesquioxide (Nd_2O_3) as a potential impact surrogate for americium sesquioxide ($^{241}Am_2O_3$). [21] As demonstrated by the efforts of the Leicester team, a fair bit of development is necessary to ensure that the radioisotope and non-radioactive surrogate materials behave similarly, both chemically and in response to physical insults.

10.4.3.3 Leveraging Legacy Design Concepts

Most commercial RTG design concepts estimate relatively low electrical power outputs of 10-25 W_e compared to the approximate 118 W_e BOL power output of the MMRTG. However, the expected masses of a 10 W_e ^{90}Sr RTG and a 10 W_e ^{241}Am

RTG are 40 and 10 kg (88.2 and 22.1 lbs), respectively. These high masses, when compared to an equivalent ^{238}Pu RTG system, result from the lower power density of the fuels, and for a ^{90}Sr RTG, the added shielding mass. Because of the low power density of the strontium-90 and American-241 fuels for commercial RTG concepts, it is not feasible to replicate the design of the GPHS module for encapsulating the radioisotopes and protecting them during launch and re-entry accidents. However, drawing from safety-related material or alloys from heritage or legacy flight-qualified fuel encapsulant designs is a viable design approach that could help streamline the launch approval process. In addition, this approach leverages information and data from decades of DOE and NASA safety testing campaigns.

For example, in the americium-241 heat source design shown in Figure 10.15, the University of Leicester team is employing 3D carbon-carbon materials comparable

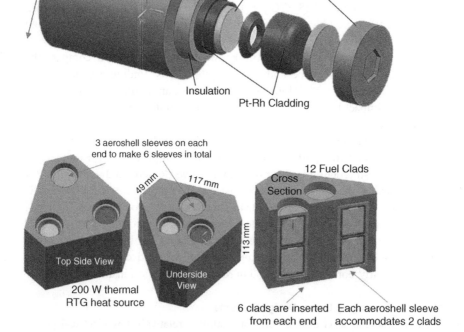

Figure 10.15 Encapsulation and heat source architectures for an ^{241}Am RTG design concept.

to fine-weave pierced-fabric (FWPF) used in the GPHS Module and an encapsulation alloy similar to the alloy in lightweight radioisotope heater unit (LWRHU).

Safety-related material and alloy selections from flight-qualified RTG systems are not the only design cues worth considering. In addition, significant levels of information from SNAP (Systems for Nuclear Auxiliary Power) legacy systems can be revealing for commercial RTG designers. Design cues, such as insulation, thermoelectric couple location, heat rejection design, heat source support, and isolation, are just some of the design cues that could be useful. Leveraging legacy system design cues can accelerate the design process for commercial RTG systems and minimize the risk of schedule delays by employing proven and tested design strategies.

References

1 The NASA Office of Technology, Policy, and Strategy. https://www.nasa.gov/offices/oct/home/roadmaps/index.html (accessed 1 December 2022).

2 Bennett, G.L. and Skrabek, E.A. (1996). Power performance of US space radioisotope thermoelectric generators. In: *Fifteenth International Conference on Thermoelectrics. Proceedings ICT '96*, 357–372. https://doi.org/10.1109/ICT.1996.553506.

3 Mason, L.S. (2007). Realistic specific power expectations for advanced radioisotope power systems. *J. Propuls. Power* 23 (5): 1075–1079. https://doi.org/10.2514/1.26444.

4 Balint, T.S. and Stevens, J. (2016). Wicked problems in space technology development at NASA. *Acta Astronaut.* 118: 96–108. https://doi.org/10.1016/j.actaastro.2015.09.019.

5 Sadin, S.R., Povinelli, F.P., and Rosen, R. (1989). The NASA technology push towards future space mission systems. *Acta Astronaut.* 20: 73–77. https://doi.org/10.1016/0094-5765(89)90054-4.

6 National Academy of Sciences (2011). *Vision and Voyages for Planetary Science in the Decade 2013–2022*, 398. National Academies Press http://www.nap.edu/download.php?record_id=13117 (accessed 1 December 2022).

7 Caillat, T., Borschevsky, A., and Fleurial, J.P. (1993). *Proceedings of the 11th International Conference on Thermoelectrics*, Arlington, TX (ed. R. Rao), 98. University of Texas, Arlington.

8 Ravi, V., Firdosy, S., Caillat, T. et al. (2008). *AIP Conf. Proc.* 969: 656.

9 Uher, C. (2021). *Thermoelectric Skutterudites*. CRC press.

10 Bennett, G.L. and Johnson, E.W. (2003). First Flights: Nuclear Power to Advance Space Exploration. *AIAA 2006-4191*, International Air & Space Symposium and Exposition, 17. https://nuke.fas.org/space/first.pdf (accessed 1 December 2022).

11 Bennett, G.L. (2006). Space Nuclear Power: Opening the Final Frontier AIAA 2006-4191. In: *4th International Energy Conversion Engineering Conference (IECEC)*, 17. http://dx.doi.org/10.2514/6.2006-4191.

12 McNutt, R.L. and Ostdiek, P.H. (2015). Nuclear Power Assessment Study–Final. *TSSD-23122*. Johns Hopkins University Applied Physics Laboratory, MD, 185. https://rps.nasa.gov/system/downloadable_items/37_NPAS_Final_Report.pdf (accessed 1 December 2022).

13 Woerner, D., Matthes, C.S.R., Hendricks, T.J., Fleurial, J.-P. et al. (2017). Next Generation Radioisotope Thermoelectric Generator Study Final Report. *JPL D-99657*. NASA Jet Propulsion Laboratory, California Institute of Technology. https://rps.nasa.gov/resources/73/next-generation-rtg-study-final-report (accessed 1 December 2022).

14 Overy, R.D., Sadler, G.G., Fleurial, J.P., and Hall, D.G. (2021). Radioisotope Power for Scientific Exploration. 52nd Lunar and Planetary Science Conference (LPSC 52). http://www.hou.usra.edu/meetings/lpsc2021/pdf/2770.pdf (accessed 1 December 2022).

15 GPHS-RTGs in Support of the Cassini RTG Program (1998). Final Technical Report. Lockheed Martin Astronautics, PA, 481. https://doi.org/10.2172/296824

16 Harmon, B.A. and Bohne, W.A. (2007). A look back at assembly and test of the new horizons radioisotope power system. *AIP Conf. Proc.* 880: 339. https://doi.org/10.1063/1.2437472.

17 Cheikh, D., Hogan, B.E., Vo, T. et al. (2018). Praseodymium telluride: a high-temperature, High-ZT thermoelectric material. *Joule* 2 (4): 698–709. https://doi.org/10.1016/j.joule.2018.01.013.

18 Weinzierl, M. and Sarang, M. (2021). The commercial space age is here. *Harvard Business Review* (12 February) https://hbr.org/2021/02/the-commercial-space-age-is-here (accessed 1 December 2022).

19 NASA Commercial Lunar Payload Services (CLPS). https://www.nasa.gov/subject/16440/commercial-lunar-payload-services-clps/ (accessed 1 December 2022).

20 National Aeronautics and Space Administration (2019). *Announcement of Opportunity Discovery 2019 - NNH19ZDA010O*, 56. Washington, DC: NASA Science Mission Directorate (SMD) https://nspires.nasaprs.com/external/solicitations/summary!init.do?solId={1A1E6E89-A0C7-0034-0802-ED6E3C8C7E49}&path=open (accessed 1 December 2022).

21 Ambrosi, R.M. et al. (2019). European Radioisotope Thermoelectric Generators (RTGs) and Radioisotope Heater Units (RHUs) for Space Science and Exploration. *Space Sci. Rev.* 215: 55. https://doi.org/10.1007/s11214-019-0623-9.

22 Hong, J., Son, K.-J., Kim, J.-B., and Kim, J.-J. (2020). *Development of Engineering Qualification Model of a Small ETG for Launch Environmental Test*. TN: Nuclear and Emerging Technologies for Space. https://nets2020.ornl.gov (accessed 1 December 2022).

23 Conca, J. (2021). Zeno power: clean Plug-And-Play power anywhere in the Universe. *Forbes*. https://www.forbes.com/sites/jamesconca/2021/12/09/zeno-power- providing-clean-power-in-all-the-hard-places/?sh=684c352b1485 (accessed 1 December 2022).

24 National Aeronautics and Space Administration (2020). NASA's Planetary Science Portfolio – Report IG-20-023. NASA Office of Inspector General, Washington, DC, 12. https://oig.nasa.gov/docs/IG-20-023.pdf (accessed 1 December 2022).

25 Mason, L.S. (2007). Realistic Specific Power Expectations for Advanced Radioisotope Power Systems. *J. Propuls. Power* 23 (5): 1075–1079. https://doi.org/10.2514/1.26444.

26 Matthews, J. (2021). Fuel design and shielding design for radioisotope thermoelectric generators. US Patent 202, 103, 910, 92A1 filed 16 December 2020, and published 16 December 2021.

27 National Security Presidential Memorandum on the Launch of Spacecraft Containing Space Nuclear Systems/NSPM-20 (2019). https://trumpwhitehouse.archives.gov/presidential-actions/presidential-memorandum-launch-spacecraft-containing-space-nuclear-systems/ (accessed 1 December 2022).

28 National Aeronautics and Space Administration (2022). *Nuclear Flight Safety - NPR 8715.26*. Washington, DC: NASA Office of Safety and Mission Assurance https://nodis3.gsfc.nasa.gov (accessed 1 December 2022).

29 International Atomic Energy Agency (2018). Specific Safety Requirements No. SSR–6, Regulations for the Safe Transport of Radioactive Material. https://nucleus-apps.iaea.org/nss-oui/collections/publishedcollections (accessed 1 December 2022).

30 Buenconsejo, R.S., Lal, B., Howieson, S.V. et al. (2019). Launch Approval Processes for the Space Nuclear Power and Propulsion Enterprise. *D-10910*. IDA Science & Technology Policy Institute, Washington, DC. https://www.ida.org/research-and-publications/publications/all/l/la/launch-approval-processes-for-the-space-nuclear-power-and-propulsion-enterprise (accessed 1 December 2022).

31 Cacuc, D.C. (ed.) (2010). *Handbook of Nuclear Engineering*. Springer Nature https://doi.org/10.1007/978-0-387-98149-9.

32 Lipinski, R.J., James, B.T., Bignell, J.L., Bixler, N.E. et al. (2014). Launch Safety Analysis for Radioisotope Power Systems. *SAND2014-4708C. Institute of Nuclear Materials Management 55th Annual Meeting*, Atlanta, GA. https://www.osti.gov/servlets/purl/1146717 (accessed 1 December 2022).

33 Cull, T.A. (1989). General-Purpose Heat Source Development: Extended Series Test Program Large Fragment Tests. *LA-11597-MS*. Los Alamos National Laboratory, NM. https://doi.org/10.2172/5857636

Index

Page numbers referring to figures are *italic* and those referring to tables are **bold**.

The Technology of Discovery: Radioisotope Thermoelectric Generators and Thermoelectric Technologies for Space Exploration, First Edition. Edited by David Friedrich Woerner.

© 2023 John Wiley & Sons, Inc. Published 2023 by John Wiley & Sons, Inc.

Printed and bound by CPI Group (UK) Ltd, Croydon, CR0 4YY

16/04/2025

14658583-0002